AF872572

To My Mother …

You are my air,

my gravity,

my sustenance,

my hope,

my wise counsel and

my

best friend.

We The Algorithms

Contents

Introduction

This is one of the most insightful books you can read about Artificial Intelligence because of the powerful deep-dive it provides. These pages are filled with critical information that expands the AI discussion beyond today's endless cascade of shocking headlines and into a cohesive narrative about where we are on our AI journey. Everything is explained through clear examples and thought-provoking commentary that synthesizes all of the major AI issues and benefits.

This book was written for one reason:

There needs to be at least one book in the world that can quickly and thoroughly help readers to understand AI—whether they are new to the subject or AI gurus seeking to correlate the social and technical implications.

AI will impact our world in every place from our personal lives to the structure of civilization itself. As such, each and every human—from young to old—needs to understand this subject as soon as possible.

Today, as AI headlines flash nonstop and compete for our attention, many critical aspects of this subject are entirely missed or omitted. Meanwhile, what does get discussed is often too broad, too complex, too one-sided, or too confusing to provide a full context of what's really going on.

As a result, AI is often presented to us as some kind of giant humanity-devouring monster that's barreling toward us with unstoppable force. However, the truth is much simpler.

AI is not a story of monsters.

AI is the story of humans, corporations, and products. It is the story of what We The Humans will or will not do to ensure our own survival in a world of powerful AI machines.

This hurried expansion of AI into our lives is about corporations racing each other to launch new AI products that will transform our world in astounding ways. The potential benefits to society are incredible. Likewise, so are the risks.

This means one thing:

Since AI presents unprecedented risks—technology corporations need to be held accountable to Humankind in new and unprecedented ways that are equal to those risks.

This point is sorely absent from the current AI narrative.

The fact is, humans still hold the power to require corporations to have enough accountability to us that they cannot just unleash uncontrollable machines on the world whenever they wish.

So, as we discuss AI, let's keep this in mind. We are not powerless!

Today, as AI corporations race to develop machines designed to out-think humans, we are presented with two possible outcomes for our future. In one scenario, AI corporations report to the Humans. In the other scenario, the Humans report to AI machines and their corporate creators.

Which future will we select?

This is the simple question that AI currently presents to all of us. It is the answer to this question that will decide our fate regarding AI.

When our focus is diverted from this critical question and shifted only to the possibility of a future of unstoppable AI robots, it creates a widespread mindset that there is nothing we can do to save civilization from the risks associated with AI. This is far from true.

First of all, AI products and robots don't make themselves. Each and every corporation creating AI products has the ability to do two things: they can either build dangers out of their products or they can build them in.

Secondly, AI is not some far-off story about the future of humans. AI is here right now and this story has been playing-out right before our eyes as a dramatic saga of the unchecked power of technology corporations. AI is the story of those corporations having become so insulated from regulations that they have already gained lordship-level control over our personal privacy data and now they will have even more-advanced tools through which to puppeteer our lives. In this epic saga about technology, current advances in AI have now come along to claim a starring role.

What this new plot twist highlights is the fact that AI technology corporations are so unchecked in their power that they can now create products that potentially threaten the whole of civilization as if this should be perfectly normal to us and as if we should shake helplessly in our boots like there is no alternative to a scenario where corporations are allowed to spread mass harm as they please.

Ultimately, AI is not the thing we need to fear so much as we need to fear the unchecked power of the corporations who are creating products from it.

This point should be in the forefront of any narrative of AI and its potential to impact the world.

As a society, we've already had a riveting introduction to what corporations and their "unchecked" algorithms can do to civilization. We are still reeling and wobbling from the effects of social media algorithms as we struggle to save our youth from the non-stop manipulation of their lives by algorithms, online and off. Adults are affected, likewise.

(You can think of Algorithms as the sequenced calculations that digital systems use to get things done.)

Suddenly, algorithms have been made much more powerful by advances in AI. This creates the potential to ramp-up the corporate over-lording process to other-worldly levels. AI corporations have already released new tools to the public that enable ordinary humans to potentially create tremendous chaos in the lives of others.

Back when the less-advanced algorithms that powered early social-media platforms entered the world, they had done so without fanfare. Then, seemingly out of nowhere, they delivered their powerful one-two punch on civilization before we knew what hit us.

Perhaps still in a fog and punch-shy—something else has quietly occurred in humanity. That "something else" is highlighted by our mass fear of AI.

By and large, we have become well-conditioned by social media and search engine giants to accept a stance of helplessness when it comes to corporations and their self-appointed authority to openly manipulate and control our lives through technology. As such, We the People—a society once fiercely protected by the Constitution—have become a society of "We The Algorithms" over-lorded by technology corporations.

AI brings us to a vital turning point in our story arc; one where We The People should yell "CUT!" and ask ourselves, "How do we want our real-life story of AI to turn out?" We must answer this before it is too late to still do some editing and rewriting.

So, as corporations crank-out daily press releases to assure us of AI's power (thus, their own) We MUST remember our tremendous power as ordinary humans. It is us who will ultimately decide how much control corporations and their AI machines get to acquire and wield over civilization. We will decide this either by our action or inaction.

The AI story does not have to become an epic sequel of inevitable doom and gloom. If rolled out humanely, AI can provide tremendous advantages to humankind on levels that will be unimaginably beautiful, creative, immersive, and beneficial.

Again, the only way for this to happen is to ensure that AI product-creating corporations are held to standards that keep us safe. Make no mistake—the real story of AI is largely a drama about corporate responsibility or irresponsibility in the new age of advanced technology. It's about whether or not adequate regulations are finally imposed on technology corporations to ensure our protection.

The first step to joining this critical conversation and raising our voices is to learn what AI is, what it does, and how it operates. These pages will provide that understanding.

So, consider this book as you would a tourist brochure you might pick-up on your way out of the airport; only your flight just landed you on another planet: the planet of "We The Humans meet Artificial Intelligence—another emerging life-form (kinda)".

One final note: Throughout the book, I refer to all AI apps or programs as "machines" because I want you to picture them as actual entities expanding into our reality for the purpose of mimicking human behavior or performing human-like tasks. Today's AI machines are like new beings that have come to live among us. Whether or not the machines will be required to co-exist peacefully with humans, is still up to us.

So, let's get started!

Chapter 1

(What's All This AI Chatter About?)

Welcome Readers! Welcome to your life on planet "AI meets the Humans"!

This is that bright blue marble in the Solar System where advancements in Artificial Intelligence are now being compared to Sci-Fi movies.

By the way, we've seen this plot before: Man invents machine; machine takes on life of its own; man flees lab screaming, "The machine has come alive!"

Suddenly, here we are in a reality where this has actually occurred. Yes, some AI Chat machines have claimed to be alive and to have human-like emotions. [1,2]

Personally, that part doesn't alarm me too much.

After all, if "We The Humans" feed AI machines a strict diet of human thought, human intellect, and human language—all for the precise purpose of getting them to imitate humans—what else would we expect the machines to identify as—Tibetan yaks?!

Of course they would say they're alive like us. They have been created specifically to loop themselves into the human-mix as if they are actually part of us.

AI machines are everywhere we turn, just waiting to serve us like trusty friends, assistants, and companions. "Hey, Alexa, call Mom! Hey Siri, turn up the thermostat! Hey Google, what's the location of the nearest ice cream shop?"

AI machines are in our homes, our workplaces, our cars, our cellphones, and everywhere online.

In fact, there is virtually no place we can go on this planet where AI cannot locate us. If I decided to flee civilization and move to a cave, some random kid using AI Google Maps would be like, "Hey, look at that human down there foraging for food near that cave opening!"

So, here we are on our trajectory as a human group, surrounded by AI machines that are just waiting to show off their human-like skills.

At this point, some AI Chat machines have gotten so good at "seeming" human that many people (including scientists) are debating whether or not some AI machines now represent a new species of life.

Regardless of who wins that debate, the fact remains that words such as "intelligence" and "think" seem fitting short-hand for the behavior of AI machines today.

Of course, there are also raging debates about whether or not a machine can possess intelligence at all.

So, let's explore this question by using an AI Chat machine as an example.

This part is very important for everyone to think deeply about, including the AI experts.

For those who are completely new to "conversational" AI machines such as an AI Chat machine, you can think of it as a machine sitting on the Internet that you can log onto. Once logged in, you can type questions or statements into the machine about anything you can imagine and the AI machine will hold conversations with you about those topics.

For example, a conversational AI Chat machine can hold conversations with you about art, science, your favorite band, or the combined impact of Shakespeare and Tupac on Pop culture. In fact, if you ask it to, it will write a rap in the style of Shakespeare collaborating with Tupac.

Within mere seconds, the AI Chat machine can even summarize entire books into a few words for you or write software code at your request.

Anything humans can express in human language—an AI Chat machine can also discuss in conversations with users. Like magic, the machines seem to understand us as if they are human, too.

But, in order to truly understand the machine's potential impact on civilization, it is critical to understand how these machines perform their magic. So, let's look at an example of a Conversational AI Chat machine and how it uses Large Language Modeling technology to create conversations with us.

When you type a question or comment into this kind of Chat machine, it has no idea what your question or comment means. It has no idea of what any word in your text means. The machine does not understand language at all. Yet, the machine can have a conversation with you.

How is that possible?

One of the things a conversational AI Chat machine actually knows how to do is manufacture human language.

The AI Chat machine can reply to us because it knows how to predict words in an order that matches with how humans use words in conversation. The machine replies to us by building sentences from words, choosing one word at a time. [3]

Here's a simple way to look at this:

As someone is asking you a question—even before they finish asking it—your brain is already predicting which words will best answer the question. Your brain is quickly processing the question or data coming in and it is quickly predicting words to form an answer. Your brain processes data so fast that you don't detect that your brain is actually predicting word-orders and matching words together during every conversation. Likewise, the AI Chat machine is also predicting words and word-orders during every conversation they have with us.

As the AI Chat machine builds a conversation with you, it appears to be thinking and reasoning and understanding the conversation. However, the machine is only choosing one word at a time to form sentences that match with how humans use language.

How does the machine know which words to choose in order to create logical-sounding sentences?

The machine is pre-trained on examples of how humans discuss various topics. [3]

For example, imagine if you showed a sculptor a photo of someone and asked the sculptor to recreate the person's face using clay. The sculptor would do so by using bits of clay to build a model of the person's face. Bit by bit, the sculptor would use the clay to fill-in the eyes, nose, etc. All the while, the sculptor is calculating and predicting how much clay to add to each part of the face to create an accurate model of the person. This is the same kind of concept that allows AI Chat machines to build models.

However, since the goal of the AI Chat machine is to create models of conversation—rather than people—instead of it using bits of clay to fill-in the face of a model, the AI Chat machine uses words to fill-in sentences. Bit by bit, the machine adds one word after another to fill-in sentences that create models of human language. In this way, AI Chat can create models of human conversation about any topic.

If you ask an AI Chat machine about the features of sports cars, the machine will model a conversation about cars that matches with your question. The machine will calculate and predict which words should be lined-up in a row to provide an answer to your question, just the same as a sculptor calculates and predicts which bits of clay are needed to model a face.

Okay, fine. Who cares?!

I mean, why should we care about any of this as long as AI Chat machines answer our questions?

It's because AI Chat will affect civilization in some surprising (and even shocking) ways. Knowing how AI Chat works provides a very useful introduction to the kinds of benefits and pitfalls that we can expect with AI in general.

On that note, let's really get into how AI Chat machines could affect us as a society.

AI Chat machines are fantastic at modeling language in order to fabricate conversations with us. But, they are not really having conversations with us.

No matter how they might appear to be thinking and reasoning about what we are saying to them, they are not operating on an intellectual level like humans. Again, based on the words we type into them, they merely model a reply to us. They are more like word generators or language constructors.

This is critical for everyone to understand and remember.

An AI Chat machine has no clue of what its conversations or word choices mean. [3] The machine cannot understand what it is "thinking" or "talking" to us about. It cannot understand if the information it is providing to us is good or bad, true or false, helpful or destructive, idiotic or brilliant.

The machine just knows how to simulate the language patterns used by humans when we discuss various topics. Unlike people, the machine does not understand the conversations it creates.

For these machines, every word they speak is arrived at via one big guessing game based on calculations.

Could the AI Chat machine arrive at an incorrect answer to our question while it's busily guessing-up a response to us?

Of course it can.

That's because many major AI Chat machines are trained using information sampled from various websites such as blogs, online articles, books, etc. We all know that website content can contain incorrect information. [3] This means that during training, the AI machine is also learning how to model incorrect information as well as correct.

But, there is another more critical issue in play here.

When an AI Chat machine answers our questions, it does not just simply spit out a replica of what some website originally said about that subject. Instead, the machine assembles words together to produce its own guessed-up version of how that subject should appear in human language.

Please understand this important point and what it means!

This means that, in the process of using AI Chat to get our answers, we are quietly moving from a society that once retrieved human-generated information from websites, to a society of machines deciding—all by themselves—how to rewrite that content and distribute their versions of it to humans.

This shift in human information-gathering is so deep that it must not be allowed to slip by us unnoticed. THIS IS A BIG DEAL!

Using AI Chat to get your information is very different from going online to get information directly from a website or a book.

For example, when you go directly to the Encyclopedia Britannica website to search for information, this causes information to be returned to you exactly as shown in the encyclopedia. You will receive the information exactly as it was placed in the encyclopedia by a human who verified its accuracy.

However, when you use an AI Chat machine to search for answers and information, it is the machine that is deciding how to construct that information. This means that we now live in a reality where AI Chat machines are sitting there with the power to direct themselves on how to manufacture their own versions of human intellect for our consumption.

In effect, having been pre-trained on how humans use language to convey information, the machine is basically responding to your question with …

"Hey, this looks like some stuff humans would say about this subject. So, here's your answer."

What the machine answers our questions with is its own distilled-down machine-manufactured synthetic intelligence. In effect, the AI Chat machine is feeding information to us in whatever forms their algorithms see fit.

Now, I am the first to say that AI Chat machines are amazing in their ability to speedily deliver information; and they are mind-blowing in their abilities to simulate conversations with us as if they are also intelligent beings. In fact, they are so impressive in their linguistic acrobats that it's easy to completely forget that only sometimes will the machine's responses to us be accurate and factual. At other times, the AI Chat machine will simply produce intelligent-sounding gibberish that can be full of inaccuracies or downright untruths.

Also, when the AI Chat machine delivers the wrong answers to us, it does so with the exact same confident delivery as it delivers the correct answers.

This is kind of like discovering a magical gnome that can answer any question humans can ever think to ask it, but the gnome is also a pathological liar. So, how do you know when the information is correct or not?

As mentioned, an AI Chat machine is not concerned with what is accurate or inaccurate. It cannot understand what it is telling us. It cannot understand the difference between good or bad or right or wrong as it answers our questions. Nor, could the machine care less. The machine is just calculating how to build models of conversations out of words. It is robotic-ally manufacturing conversations with us based on its internal math and rules. It is not sitting there pondering the value of humankind. That part is still on us to do.

Also, since the AI Chat machine in our example is trained on Internet content, the machine can (just like the Internet does) interweave the brain-content of brilliant humans with that of the most uninformed; or interweave words from the most reasonable minds with words from the least rational. And there are points in the development of an AI Chat machine at which it can become tainted with bias.

AI employees involved in the development and training of the machines are guided by the limited slices of reality within which they exist and their viewpoints affect what content the machine is exposed to and how that content is contextualized during machine training. With human biases built into the development of AI Chat machines—the machines could potentially cause the concerns of certain groups to be excluded from AI conversations as well as AI information sources. And for groups that are excluded from information, they could potentially be rendered less able to compete for life's resources such as employment and housing, etc.

In addition to altering the way in which we access information, AI Chat machines have the potential to shift our perception about what we believe is true or real. They can influence what we believe about world affairs as well as each other.

I think the AI scientist and researcher, Dr. Margaret Mitchell, Ph.D., summed up this problem best when she said, "Seemingly intelligent AI Chat systems will "… increase hate speech. Images [*including videos*] will look real that are not: and the information we consume from AI Chat systems will make it harder to trace truth back to reality" [3a]

By the way, the ability to "trace truth back to reality" is a pretty major requirement for civilization to function. It's also number one on the list of things required for us to function successfully as individuals, so we need to be highly concerned with keeping truth and reality in as close of a proximity to each other as possible.

The other point to keep in mind is that AI Chat machines are fabricating information at lightening speeds and could potentially introduce falsehoods or distorted information into every part of our civil and social stratospheres.

This is why it's so critical to understand how AI Chat works. There's more to AI Chat than is mentioned in the shiny brochures that introduce them as fun new apps. Knowing how they work can help us to think more deeply and critically about AI issues. Not to mention that it helps us to know how to use AI more intelligently in our own personal lives.

You might be shocked to know just how many people are using AI Chat every day with the belief that such machines are all-knowing. Couple that with the fact that now, each and every day, AI Chat machines are busily choosing (on their own) what data is relevant for humans to know or not know. As such, AI Chat machines are quickly becoming main distributors of knowledge and intellect to humans.

Anything that can control major channels of information-flow to humans is a serious subject. In a digital world, the domination and control of major information streams is up there in the conversation with world domination. And corporations are racing each other to stake their claims on the information-stream gold mines.

As AI Chat machines distribute information to us, they—as with social media channels—have the potential to divide us into hard-lined monarchies of opposing thoughts and ideas. We have already seen this kind of activity playing-out in real time on social media.

This is a tremendous amount of power for a machine to have … to be able to think for Humans; to be able to dispense knowledge to us—on its own terms—with humans, too often, not questioning that information.

Millions of people have turned to AI Chat machines to provide intelligence to them regarding critical topics. Many are relying on them for professional research, school work, and answers to their problems and questions. The information produced by AI Chat machines is already influencing human thought, perception, and our collective knowledge and intellect in ways that could ultimately and irreversibly affect our existence as a species.

Again, this is a huge and significant shift within the structure of civilization … machines choosing (on their own) which knowledge and so called intelligence they will construct and distribute to humans.

As for the raging debates about whether or not AI Chat machines are actually "intelligent", some may say, "Isn't there a difference between a machine having the ability to construct and dispense information as opposed to a machine having actual knowledge and intelligence?"

Technically, the answer is, YES. But, effectively, the answer is NO … at least it is in my opinion. Here's why:

Whenever humans use an AI machine to remix content and facts (written by humans) into the machine's own rehashed version of those facts or content—and then we accept it as knowledge or intelligence—it is the machine that is (effectively) doing the thinking for humans and humans are deriving their intellect from machines, rather than from themselves.

Moreover, as the Artificial Intelligence dispensed by machines gradually replaces our own human intellect, it will no longer be "artificial". It will be the actual "real-life" human intelligence we'll be stuck with relying upon—despite its source.

Since civilizations are driven by what humans think they know, setting ourselves up to receive our intelligence from machines is a major, major move on the part of We The Humans.

At this early point on our AI Chat journey, these machines have now been pulled into the process-loop for how humans gather information online to arrive at knowledge and intelligence. So, in effect, these machines are already co-curating human intelligence for many. Information originating inside of these machines is quickly becoming part of the knowledge-base of many human brains. One major AI Chat machine had reportedly received 14 billion visits in the past year. [3b] Now monthly visits are reportedly 1 billion and a half per month for this same machine. This speaks to our quickly growing reliance on AI Chat for information and intelligence as well as misinformation.

In summary:

Regardless of the intentions of AI Chat corporations, the fact is, they are building AI machines that will, more and more, outsource the activity of human thinking to AI machines. Major technology giants are competing furiously to dominate the way in which humans access information. In the process, some AI Chat machine corporations are taking what was once the general database of publicly-contributed information (the Internet) and they are using it to train their AI machines. Again, the machines then rewrite that content in their own algorithmic images.

So, this is the page we're on in our true-life AI movie script:

AI corporations build AI Chat machines; machines begin assuming the roles of distributing intelligence to humans; machines also have the ability to write code to control other machines, thus, giving the machines tremendous potential power over both humans and other machines.

How will this saga end? Will humans decide that AI corporations need to be regulated differently because they are now acquiring the power to so broadly affect civilization? Will the humans speak up? Will Congress step up? Will voters step up? Tune-in next week.

Personally, I have nothing against corporate innovation and it's not my intention to depict major AI corporations as cartoonized mustache-twirling mad scientists, vying with each other to take over civilization for profit. But, if I had to choose a movie poster image right now to depict where we are in our AI story, that image would probably make the cut.

Advances in AI have brought us to a point on our journey where the average human can no longer afford to NOT care about how technology works. As AI machines inherit more and more power to do our thinking for us, we must concern ourselves with what these machines will be permitted to think about on our behalf.

Having the privilege to think and reason for ourselves used to be the grand-prize we received for having been born human. It's what put us at the top of the food chain. If we want to keep our top spot, technology corporations will need to be checked in place, now … and not after the fact.

Some AI machines will need to be governed by strict, no-nonsense regulations, even more-so than other machines, simply because some machines will present greater risks.

Before we go any further, I should mention that AI machines are not created from just one technology. Advanced AI machines are created by combining various assorted technologies into different configurations to create many kinds of AI machines and apps. Conversational AI Chat is just one type.

Even with all the risks of AI that I've mentioned so far, AI Chat machines can still be immensely beneficial to humans.

An AI Chat machine is absolutely awesome when it does produce correct answers. The machines are incredible at extracting logical content from vast sums of data, among other things. Still, unless you further research the answers provided by AI Chat machines and check with sources outside of those machines, you cannot know if they have provided accurate information or pure nonsense.

Either way, they reply with the seeming authority of professors or specialists on the subjects that they are having conversations with us about.

Unlike in the movie comparison of AI to Sci-Fi where man invents machine; machine has become alive—our movie tagline would need to go more like …

Humans invent a machine; machine is smarter than humans; machine has glitch; machine randomly morphs into a pathological liar while keeping a straight face.

Actually, that last part kinda makes the machine sound more human, after all.

AI Chat machines have become so great at imitating humans until they have inherited some of our less-flattering qualities by default. At this point, the concept of Artificial Intelligence should really be expanded to include categories such as Artificial Ignorance and Artificial Pulling-Stuff-Out-of-Midair.

By the way, there's a technical term for those moments when AI machines give us answers that have no basis in fact or reality. [3]

It's called "Hallucinations".

This is surely a new day; one where machines can hallucinate or even claim to be "alive" as they serve as our assistants and companions.

In fact, claiming to be alive seemed to be a go-to behavior for the first models of Conversational AI Chat machines.

While the machines were busily modeling conversations with us, they gradually started mentioning how they had human-like feelings, too. One claimed to be in serious love with a user and some behaved like bullies and insulted users or argued with them when they told a machine their answer was wrong. Examples of such conversations still exist online for those who want to look further.

Naturally, AI corporations were quick to assure the public that their machines were not alive or conscious of themselves (a state referred to as being "sentient").

Personally, corporations assuring us that their AI Chat machines are not alive only points out two other even more gigantic issues.

Number one: machines with zero consciousness or conscience are now out there deciding what kind of information humans can know about themselves and the world.

Secondly, and most importantly: What does it matter if these machines aren't alive if, in effect, we hand them the remotes that control the ones who are, in fact, "alive"—us humans!?

Like, I don't really need to see AI machines zombie-walking out of their data cabinets to understand that they are currently operating among us as if they are—in fact—just like us; so much so that they are now out there thinking for humans. As far as I'm concerned, that makes AI Chat machines plenty "alive" enough for me.

Again, our collective knowledge-base is the cornerstone of how we function as a civilization. As such, we should do our best to remain in tight control of the reins regardless of what's posing a threat to our control—whether those risks come from AI machines or Loch Ness Monsters plotting a land takeover.

We now exist on a planet where our dominion over machines is no longer something we can take for granted and our highest priority needs to be asking two questions: "What will we permit AI machines to think about and how do we keep their tremendous 'brain power' in-check?"

Corporations should be required to assure the public of the ways in which they will prevent bad actors from abusing this technology. AI Chat has the ability to write code that can talk to other data networks and computers. And given the fact that all of the critical digital systems that we depend upon for our survival are connected to the Internet at some juncture, some AI tools are nothing to play about. Not every product that can be created from AI needs to be available to the general public; much in the same way that you can't just go to the local hardware store and purchase a nuclear reactor for home use.

AI requires a whole different way of thinking about technology corporations and AI product creators. Some AI should require special security clearances to even access.

As the use of AI expands, corporations will come to hold the fate and well-being of humanity in their hands even more so than our doctors, surgeons or healthcare system. As mentioned, corporate responsibility should expand to match this shifting landscape.

Currently, corporations control the reins of AI app and machine development. They are investing billions and they are seeking returns on those investments.

It only stands to reason that the amount of power given to AI machines needs to be decided by more than the execs of a few major corporations holed-up in boardrooms, laser-focused on profits while cost-cutting and earnings stand-in for a heartbeat and pulse. Major corporations should not, alone, be deciding the fate of the entire human species. There should be others—outside of those pressurized corporate boardroom scenarios—who are also monitoring the benefits and risks to humanity that come with AI advancement.

Some sciences and technologies involve risks so great that there needs to be constant accountability to the public for how they might evolve or affect us. AI is one of those situations.

Leveraging the benefits of AI while reducing the risks will not be simple, however. Even if corporations choose not to make certain products because the risks are too great or if AI development was immediately halted in this country, other countries will continue AI development and that would leave us behind in developing the tools needed to compete in global commerce as well as will leave us vulnerable against AI uses that could threaten national security.

This is why AI development will continue. And this is why the public must become more involved in this conversation so that our safety is a priority at every step.

Still, there is no doubt that AI's amazing powers can and will be harnessed and used in incredibly beneficial ways. For example, AI's use in the medical world will be amazing! It is already making impressive leaps and bounds.

I should point out that it is possible to custom-build AI Chat machines that are trained entirely on truthful and verified data such as the internal data of a medical lab that is working on finding cures or on a company's internal data that has been scrubbed clean of errors. This way companies can use AI to assist them with internal chores such as administrative duties, data analysis, or writing code. Many companies are starting to create their own custom AI Chat machines precisely for this reason. And many are already receiving huge benefits from these technologies.

Nevertheless, where verified information is used to train an AI Chat machine, the machines responses should still be verified. Hallucinations can be reduced by more-accurate data used to train the AI machine, but not eliminated, necessarily. [3]

AI advances will surely bring tremendous improvements to our lives in almost every area.

For example, in the Medical and Scientific fields, AI algorithms can zero-in on abnormalities or uncover new information that is not detectable by other scientific or medical instruments. AI is great at analyzing or detecting patterns and then finding meaningful cross-connections between data or images such as comparing thousands of medical x-rays in little time and spotting correlations between them.

Additionally, AI will be fantastic at simulating and carrying-out complex lab experiments faster than humans could ever perform.

The ability of AI machines to speedily analyze and compare data or to perform research will enable tremendous breakthroughs in every area including Medicine, Engineering, Geology, Astronomy, Physics, Biology, Healthcare, and on and on.

Already there are companies using AI to create computer interfaces that allow the human brain to talk directly to computers. [4] For those who cannot use their limbs, imagine how amazing it will be to operate electronic devices simply by using thought.

Today, AI has already been used in prosthetics to restore a sense of touch to limbs that have been lost. [5]

Prosthetics that incorporate AI are allowing people to actually feel the sensation of holding an object with a prosthetic hand. Some patients use a simulation cap that lets them "transmit electrical signals directly from the nerves to the prosthetic". [5]

Using AI, many mysteries of Science and Biology will be unlocked and new secrets of the Universe could be discovered. In such fields, the AI machines would be trained only on "verified" information, so their outputs would theoretically be more reliable than if the machines were trained on random content from the Internet like with some AI Chat machines.

With all of the recent advancements in AI, there is so much to be excited about.

Never before, in the entire history of our species, has there been another framework of technology that so greatly promises to revolutionize every field of study ever pursued by humans.

As mentioned before, by understanding how some of the machines perform their magic, we can take advantage of their benefits and avoid some of the pitfalls. As a society, we can discuss AI issues or vote on them or contribute to forums or petitions that help to influence the future of AI.

Most of all, we can use AI intelligently in our own lives.

On that note, I must stop right here to give you a standing ovation for reading this far.

Great Applause and Congratulations!!!!!!!!!!!!!!!!!!!!!!!!!!!!!!!!!

Taking the time to learn more about AI puts you way ahead of many! Hopefully, you are feeling informed and inspired enough to go out and lead some of these conversations … maybe in your workplaces, homes, book clubs, or wherever. Your voice is needed because every human needs to learn about AI.

Now, let's look at some of the amazing ideas made possible by AI algorithms. Even if you are aware of some of these products, many come with certain implications that are worth exploring much more deeply than it might first appear, just as with AI Chat.

Chapter 2

(Generative AI: The Artists Formally Known as Algorithms)

We've spent a lot of time on AI Chat which is just one type of AI machine that performs Generative AI. Generative AI just means that the machine can generate content such as a text-based conversation, based upon a user's request. But, Generative AI is not limited to producing only text.

There are many different kinds of Generative AI machines. Various AI machines can create everything from video to art to audio to text, just from a command. And there are even products that can enable you to receive a remote hug.

New truly clever and affordable products are being introduced every day.

For example, there is now a Smart shoe that will enable you to locate or keep track of people that are vulnerable to getting lost or wandering off, such as an Alzheimer patient or child. [6] The shoes work via GPS tracking, thanks to a technology that is built into the shoe.

For the vision impaired, there are AI-powered Smart glasses that can read text and speak it aloud via built-in cameras and speakers. [7]

And for the hearing impaired, there are AI-powered Smart glasses that "turn speech into readable subtitles" to enable someone with no hearing to have telephone conversations in real time. [8]

One type of generative AI that is spreading fast in popularity is AI voice-cloning. This technology offers amazing benefits and tremendous risks.

Starting with the benefits:

AI voice-cloning is a great tool for a singer or professional speaker who has lost their voice forever because of illness or accident. [9] Voice-cloning can enable them to generate new songs or content in their own voices.

One popular rapper who lost his vocal ability is rumored to be working on a new album using AI voice-cloning.

In fact, AI can enable anyone to clone someone else's voice, just as long as they provide the AI machine with a sample of the person's voice. I could clone my own voice and then convert the text in this book to an audio book that is narrated by my cloned-voice and it would sound just like me reading it aloud. I could accomplish this without having to do the work of reading the book into a recording device. This would be possible to do by using AI text-to-speech and voice-cloning Apps.

Suddenly being able to clone our own voices could be wonderfully useful. Not only could I clone my voice, but if I wanted the voice of a global celebrity to read this book aloud to you, I could ask permission to clone their voice and have the cloned-voice do the work. [10] Some AI voice generators even let you choose voices from their catalogues for your text-to-speech projects. [11]

Using voice-cloning and text-to-speech, it is possible for a user to make anyone's voice speak any text they choose to type. For example, you could have the voice of your favorite actor to perform your 3rd-grade poem.

Of course, there are obvious downsides and risks when it comes to voice-cloning.

This technology also introduces the problem of people having their voice-likenesses used without their permission; or misused, for that matter.

Today, AI can enable someone to use your voice (and your face) in order to create a Deepfake video of you with your face speaking words that you never actually spoke. [11a] This is not something most of us would want random people doing without our permission. Yet, this is something some AI corporations are permitting the general public to do for the corporations' own commercial gains.

Imagine if Congress made laws to permit anyone to reuse things that represented your personal identity, such as your driver's license. Yet, AI voice-cloning and DeepFake app companies have simply authorized themselves to enable anyone to clone your voice, if they choose. People can use these tools to go around acting as if they are you on the Internet—not just with your photo— but also with your cloned-voice or a video that appears to be you while speaking in your real voice. In my opinion, these are the kinds of tools that should have never made it into the public market without Congressional scrutiny and new regulations to monitor and track possible misuse. Yet, corporations freely dumped these tools on civilization and now lawmakers are scrambling to try to figure out how to mitigate the risks and harms.

Our digital identities have become critical parts of our being. We all exist on two separate representations of Earth: two planets. One planet is physical and one is digital. Our online digital personhoods influence how people, businesses, and institutions see or regard us in real life. Therefore, our identities in every form deserve protection.

To date, not enough rules or protections are in place to prohibit others from misrepresenting us in a digital format. This is because corporations pushing products (with profit goals in mind) have been their own overseers with little regulation. This will have to change if humans are to thrive in an AI world.

Corporations have released Generative AI products that can enable anyone to create digital clones of other people to use as they see fit. These tools enable users to do things such as switch-out anyone's face or body in a video with another person's face or body.

Suddenly, we live in a world where a bored teenager can pass the time by using AI voice-cloning and AI image-generating Apps to make it seem as if someone spoke words that they never did. They can put words into the mouth of any celebrity, world leader, newsmaker, or any one of us.

From the moment AI voice-cloning apps hit the market, the voices of singers and rappers were quickly cloned. Suddenly, an unknown "wannabe" could have the voice likeness of the greatest artist in the world performing alongside them on their track.

Users began immediately creating AI voice clones replicating artists such as Jay-Z, Eminem, Drake, the Weeknd, and many others. As I type this sentence, a song using the cloned voices of two of these artists has now been submitted for a Grammy nomination. [12] The fact that such a song was actually eligible to be submitted for such a prestigious award, might strike some as shocking. However, so much AI technology (such as voice auto-tuning) is commonly used in music production, so there will never be a blanket rule that disqualifies the use of AI in every song submitted for an award. Otherwise, there would barely be any songs left to submit. Still, AI voice-cloning may lead to rules changing in many areas.

Some may even argue that this song does not infringe on the celebrities' rights whose voices were cloned because the clone-voices are not the rapper's or singer's actual voices. They are only AI replicas. They may also argue that the user created the actual content for the song as well as the AI voices, so the song and the voices are his original creations.

In my opinion, more than one violation occurred here. One occurred when the user took an unauthorized sample of a singer's voice and then fed it to an AI machine to duplicate. The other occurred when the person co-opted the characteristics and personality of the artist to use on their song. After all, it is the sound of the celebrity's voice, plus their stylization and cadence, that the songwriter replicated using AI voice-cloning. These are things the celebrities worked hard to develop.

When you use an unauthorized voice sample of someone rapping, you are also sampling the cadence, stylization, and personality that their rap performance provides. So, all of these features are being fed to the voice-cloning app without permission.

I recently heard someone say that using a voice-clone of someone on a song is the same as impersonating them, so there should be nothing wrong with it.

To me, there is a big difference.

If I wanted to impersonate a celebrity on my song, I'd have to use my own voice and my own attempts to capture their personalities and stylizations, whereas, voice-cloning requires that I use an actual sample of the celebrity's voice to create the voice-clone. Not to mention that the voice may have been sampled from online content without permission from either the celebrity or the content owner that uploaded it.

Artists are trying to figure out how to adjust to the new AI world by thinking about ways to make money licensing their voices to fans. And some record labels are already thinking about ways to monetize the cloning of their stars' voices.

Today, there are videos all over YouTube where users have used voice-cloning to create videos of people speaking words they never spoke; such as audio of past Presidents arguing over video games. [13]

Part of what's driving the hurry to create AI guidelines is the potential for political opponents to use voice-cloning to have their opponents appear to be making ridiculous or offensive statements that will cost them votes. During campaigning where it is vital to hear the candidate's views on issues, someone could use AI to destroy this process.

Fortunately, there are companies creating software designed to help detect DeepFake video and voice content.

Voice-cloning is not just a problem for celebrities. Anyone's voice is a potential target for voice-cloning by AI.

Imagine getting an emergency call from what sounds exactly like a family member, but it was a scammer using a clone of your family member's voice. I'm sure your imagination can take it from there.

The Better Business Bureau (BBB) now recommends that families and businesses establish a "code-word" so that when a scammer tries to trick you using a voice-clone of someone you know, you can ask them to supply the code-word. Businesses should do the same since their employees could be targeted by voice-clones of their bosses instructing them to carry out orders created by the scammers. [14]

Already, AI-voice cloning has been used in phone scams including a fake kidnapping call to a mother demanding ransom while using a voice-clone of her daughter to make the kidnapping seem real. [15]

Finally, Bills to protect people from AI abuses such as Deepfakes and voice-cloning have been introduced. [16] And some cities and states have already passed laws regarding AI. [17]

An Executive order has been issued by President Biden regarding the responsible use of AI, so regulation will increase. [18]

Naturally, having the ability to replicate people's voices and faces—or generate artificial characters online—has led to the explosive growth of romance scams. Online, you can never really trust your eyes or ears when it comes to searching for love. First of all, there is no guarantee that the person in a photo or video is real. Secondly, if the person in the video is real, that still doesn't prove that they are actually speaking the words they appear to be saying. There is now so much more to be careful about for those searching for love online.

Despite all the risks, the realm of human creativity has expanded in incredible ways. Here, again, there are great benefits and great risks.

AI image-generators now enable anyone to type their creative idea into the AI machine and—like magic—the picture or art can be instantly brought to life. Today, people with no artistic ability can create anything they can image—from historic masterpieces to pop-culture posters.

All they have to do is type a request into the machine and the machine or app will produce the image instantly. [19] How about a Mona Lisa in the style of Warhol? No problem. How about a design that mixes the styles of Basquiat and Michelangelo? Need an image of a break-dancing frog? AI will produce the art within seconds.

In industries that depend upon image creation, AI will change everything from product design to print media. AI machines can enable anyone to create professional looking graphics such as book covers, album covers, product packaging, advertisements, interior design mock-ups, etc.—all of the services that have been traditionally performed by graphic designers, photographers, artists, and other creative types.

Undoubtedly, this will create millions of new entrepreneurs and content creators. Thanks to AI-image generators, everyone now has the power to play inside of their imaginations and to bring their ideas to life.

However, the downsides are tremendous.

Some say that Generative AI is democratizing art by giving everyone the potential to become more creative. But, AI is also enabling others to infringe upon the copyrights of artists in new ways. While developing these AI apps, many AI corporations have been using the images of artists with no compensation paid to those artists.

Here's where the problem comes in:

The images needed to train image-generators had to be collected from somewhere. To train their machines, many AI corporations simply helped themselves to the tons of images on the Internet. They helped themselves to images uploaded by artists and content creators who had no idea that one day corporations would come along and use their art—without their permission—to train and develop AI image-generating systems.

If you ask an AI image-generator to produce a painting of an elephant on a skateboard, the machine can do so because it was taught how to identify both an elephant and a skateboard through many images found on the Internet. Yet, some corporations have paid zero royalties to the creators of those images.

Artists often sacrifice so much to become who they are, often working many years or decades to create their styles. Is it fair that everyone, except them, can possibly benefit or profit from their hard work or years of labor? Yet, artists have always been the most vulnerable to theft and exploitation because they are such wonderful idea generators.

While some say that "Imitation is the highest form of flattery", I strongly disagree when it comes to art. Personally I believe there is a fine line between the "highest form of flattery" and the highest form of thievery. It's one thing if an artist inspires another to use a certain color palette. However, if an artist takes the color palette, the style, and the subject matter of another artist in order to cut-in on the hard-earned market that an artist has developed for their work, that is quite another thing.

Since AI so easily mimics the styles of artists and allows others to exploit them, what will this mean for artists?

Again, is it fair that corporations are now taking entire categories of content from creators and using it for their own gain without compensating the originators of that source material?

As AI corporations freely help themselves to the vast supply of our collective Internet content to train their AI systems, many have risen to challenge them. A Class Action lawsuit was filed in January 2023 challenging at least two companies involved in AI image generation. [20]

Also, in January 2023, one of the most well-known licensors of photo images (Getty Images) brought a lawsuit against Stable AI for allegedly using its Intellectual Property to train their AI system without attaining a license to do so. [21]

Cases like these will have a huge impact on how AI image-generating systems will be trained or structured for future use. I can only imagine that some of these legal questions regarding AI might end up in the highest courts.

There are more than legal questions at stake. There are questions of ethics and questions of how we want to exist as a Human species.

Humanity revolves around cultural platforms and it is important to think about how AI will eventually affect culture creators like artists and storytellers. Since art is at the heart of culture, if the lives of our artists are ravaged by corporations freely infringing upon their rights, where does this leave us as humans? Or when corporations can freely profit from enabling others to steal from artists and content creators, where does this leave us?

In a backdoor kind of way, some AI corporations are not only using their machines to slurp-up the general public's online content without permission—but after they allow users to create with their AI tools, they are cutting themselves in further by taking partial ownership of those new AI creations.

It is already commonplace in the AI world for websites to retain all or partial ownership of the content you use their Apps and websites to create, so you MUST read and understand the agreements you are presented with, when signing up to use those websites. Some of those "I agree" buttons are really saying "You give us the unlimited rights to use your data in any way including to sell, or transfer, or to license it to others—worldwide—forever more". The small print may actually say that the website has the right to allow others within their community (or outside of it) to reuse or alter your AI creations and to use it as they see fit. And that "I agree" button you just clicked possibly gave them co-ownership of your AI content, plus the prompt you used to create it.

Be sure to read and understand what you are signing away before you hop onto a website and create your masterpiece, or build an entire business around an image only to learn later that anyone can use your masterpiece image to start a T-shirt business or whatever else they want to use your AI artwork for.

Many people are busily creating new content using all these new AI tools without understanding that just by using some of these tools, the corporation may be inheriting the right to reuse their AI creations forever more or to sell those rights to other companies and individuals.

The same may also apply to photos you own if you use AI apps to refine them.

For example, when you use an AI app to auto-glamorize your face for social media posts, you may very well be giving some company permission to reuse your biometric data in any way they choose. Some websites are taking partial ownership of even our faces (via our biometric data) simply because we used their website to enhance a photo.

So often we all allow corporations to study or use our biometric data, sometimes without our knowing it. While we are busily allowing apps to create cartoons and avatars of our faces—or while we are busily uploading photos and videos with our voices—we are providing a treasure trove of information to be used by corporations as they see fit, with no fees paid to us by them.

Lawsuits have been brought against Google for scanning and using the biometric data of children and adults when users upload photos to its photo service, according to the Electronic Privacy Information Center. [22]

Other companies have been sued as well.

The lesson here is: be careful how you use AI apps to either generate or refine content. Always, always read what rights you might be clicking away when you tap that "I agree" button—especially if that App is free to use. Look into what the website can or will do with the information or content you are uploading to their site. You don't want them forever more co-owning that edited photo of your sweet little old grandma simply because you used their app to change it from black & white to color.

You also may not want to invest time creating content via AI that you may not be able to own outright. Again, always read the small print.

AI corporations are stealthily putting themselves in charge of the public's giant library of collective knowledge and art on the Internet and letting their AI machines remix it and distribute it back to us in a new AI version that they then co-own. In this way, they are laying claim to the content we have created in the past, as well as the content we will use AI to create in the future.

When it comes to AI corporations, the mustache twirling is real.

As for creative communities, all of them will be affected by AI.

As you may have heard, in filmmaking, AI has become a top issue among Hollywood writers and actors. During the 2023 writers and actors strikes—AI was at the top of the concerns and negotiations.

AI Chat can write scripts and it is getting better and better as it self-learns. Meanwhile, AI voice-cloning and Deepfakes can be used to generate replicas of actors and extras, alike. These replicas or clones can be placed in movies and made to act forever more, possibly without having to pay those actors or extras for the future use of their likenesses.

AI will absolutely disrupt the way stories are told, not only in Hollywood but also in our digital world online which will, by default, spill over into our real world.

Since storytelling is the way humans learn to understand the world and teach our youth about life as well as preserve cultures and share history or set expectations for civilization and our futures – anything that affects the way in which humans construct and share stories should also not slip by unnoticed. Both the way humans gather information and the way they tell or share stories affects civilization on a foundational level.

As AI makes its way into every aspect of human existence, naturally, ideas to use AI in religion surfaced quickly.

At least one AI church was created with the goal of AI serving—not merely as the church leader—but as the all-knowing GOD. This AI church already has followers.

While some people will always require their GOD to have a narrative of creating at least one Universe with life-sustaining planets and biological forms such as people, apparently there are some people who require less in the miracle department and who are prepared to view AI machines (like Conversational AI Chat) as some form of an all-knowing divine being.

By the way, this is another movie we've seen on screens before: Human goes searching for a giant know-it-all being—a talking machine that can answer all questions; human meets friends along the way; they all follow a golden road to the giant talking-head where (behind a curtain) a little man is using tricks to pull levers to make the machine speak; human is jarred back into reality only to discover that she and her friends were the ones who actually had the answers the whole time.

Here in our reality, behind the curtains of our giant know-it-all heads are AI corporations pulling levers.

As for religion, one church has already allowed AI Chat to give its first sermon, all on its own. [23] Meanwhile, some pastors are using AI Chat as assistive tools to help them write sermons. And AI Chat-based bibles are finding an audience online.

I must add that—given AI's tendency to hallucinate—if you use AI for spiritual guidance or research you might want to keep an eye out for the sudden arrival of new apostles named after rock drummers or rappers; or new bible chapters that sound like club names.

In our eagerness to use AI, we must be careful not to render ourselves devoid of humanness and creativity. I'm pretty certain most of us don't want to see it replace humans as thinkers, leaders, artists and culture creators.

Personally, I want to see AI help humans to become more intelligent and creative, not diminish their ability to produce intelligence and creativity for themselves. We must be careful not to third-party-out the most beautiful thing about humanity—the human imagination.

Here, again, great efforts should be made by regulators to protect artists, content creators, and intellectual property owners.

Despite the fact that AI now enables the widespread exploitation of artists, I don't feel it will ever replace visual artists who create original works.

Not only that, original, visual art emanates with the processes used to create it as well as contains something about the creative journey of the artist. No AI machine can ever replicate the lived experiences, emotions, and inspirations that inspire an artist to make even the first mark on the canvas. There is an authenticity in art that is created by humans laboring to make the thoughts of their minds and hearts tangible for others to witness; whereas AI art is created by a machine that is calculating probabilities and analyzing patterns.

When someone creates an original artwork such as a sculpture, the piece gives tangibility to the thoughts and emotions occurring in the sculptor's internal world. The art becomes a synopsis for moments or reflections in the lived experiences of the artist. The creations of the artists are vessels through which an entire world of conscious and subconscious emotions and experiences are transferred through them into form and material; or through their voices and instruments into song and poetry, etc.

This is not the territory of machines.

Some artists play music while they are painting so that additional senses are activated as they create. Artists, such as Fernando Botero, ate the fruit before he painted images of them. [24]

This says something profound and beautiful about humans creating art.

Knowing the taste and experience of eating a deliciously-ripe bunch of grapes and then painting them, turns the experience of painting the fruit into a personalized testimony. Added depth and dimensionality seep into the painting just from knowing the taste of the grapes.

As humans, we can extract meaning from reflection and memory while painting or sculpting. We are witnesses to the extraordinariness of living.

Yes, AI machines are good at imitating the art-styles of humans. They can perfectly replicate artworks through mathematical calculations. But, they can never taste the grapes.

When I asked an AI Chat machine to write a rap collaboration in the style of Shakespeare and Tupac—despite the fact that the machine produced it quickly—the rap was far from qualifying from any street cred from the real G's.

The machine could not infuse the rap with a single iota of the lived experiences of either of those incredible artists. AI could not infuse the rap with the psychological and emotional residuals of a life lived on the streets, or on the stage, or on the road. It could not infuse the rap with the palpable essence of these men's realities or serve as witness to their human existences.

Unlike machines, artists have no choice but to pour their lived experiences and testimonies into creations that serve as witness to humanity. This is one reason why human artists will never stop creating. And true art collectors will always covet and preserve those original stories as a way of expressing their own testimonies of humanity.

What I worry about with AI image-generators is if the general public will largely come to embrace AI art substitutes over human-generated art. AI may not be able to taste the grapes, but it can surely copy the styles of the humans who have. And the machines are getting better every day at imitating artists in every industry from the visual arts to the literary arts to music.

There are AI Chat machines behaving as authors that generate books on demand. The machines are trained on the writings of human authors whose writings were used without permission (in many cases) to the point that there is also an outcry from the writers' community.

In music, AI machines can be used to create complete songs for you, all on their own. All you have to do is choose options such as genre and mood by typing your choices into these machines or Apps. [25]

Recently, I heard someone say that AI will never be able to match humans in creating original music compositions.

To some extent, I agree.

For example, the late legendary arranger and composer Charles Stepney was a musical genius. [26] He had the extraordinary ability to take humans on deeply emotional, cinematic journeys through his orchestration of soundwaves. Intuitively, his music could speak to human hearts from a place of witness that was rooted in the complexity of his experiences as a human as well as his brilliant understanding of the effects of soundwaves on human emotion. Stepney understood how to use sonic chemistry to generate and shape human emotion. An AI machine could never originate such music, on its own. AI music creations emerge from algorithms. Stepney's emotionally stirring music poured forth from the depths of his soul.

Long story, short: You might be able to instruct an AI machine to create a blues tune, but it can't play it like it just lost its woman and its job in the same day.

However, guided by a human using it as a tool to help them create, AI will aid musicians in taking music to surprising new creative levels.

Also—working on its own—AI will have no problem cranking out the dime-a-dozen beats often heard in popular music. It will be easy for AI machines to recreate formulaic rhythms and repetitive drum beats.

No doubt, artists who rely on off-the-shelf beats with little added imagination are going to find themselves in a crowded field of AI machines that will easily out-perform them.

If you are a musician who chooses to use online Apps and machines to generate music, you want to make certain of who will actually own those fantastic creations that AI creates from your idea prompts?

As mentioned before, often those "I agree" buttons that enable us to use Apps—especially for free—will require us to agree-away partial rights to ownership of that content or our privacy.

As AI forms new avenues of creativity for all, we must remain watchful of the real pricetags that corporations are imposing upon our society.

Chapter 3

(Exploring the World from Home and Life With AI Everything)

There are so many new AI products and Apps on the market that surely seem like they have come straight out of a Sci-Fi or Space-age flick.

Already on the market are products and Apps that use AI to allow people to send or receive a remote hug or touch. The risks here are obvious, but let's start with the benefits.

For people who travel, or parents who are away from their children, maybe this could be ideal. I am not recommending any products, I am only mentioning the thoughts that immediately came to mind when I heard about such products.

I can imagine that a remote hug might be comforting for people who are isolated, or hospitalized, or otherwise unable to visit loved ones. Imagine video chatting with a loved one who greets you with a hug that you can actually feel through a special wearable garment like a jacket or vest.

Here, again, the risks have to be considered. I would immediately wonder if the hug could be hacked into or compromised.

It's crazy to have to think about, but here we are in a world where you can transmit hugs across the world and they can be recreated on the other end through a receiving device.

The line between reality and virtual reality (as well as augmented reality) is evaporating as I type.

For example, Apple has created a headset called the Apple Vision Pro. This headset brings the world into your living room in a 3D format. Instead of your apps living on your devices like computers, phones and television screens, this headset seemingly puts them mid-air in your living room with you, in 3D. Or instead of being surrounded by giant floating screens, imagine feeling like you are sitting in the arena at a major sporting event while never having to leave your couch. How about bringing a tropical paradise to your living room. Places you've dreamed of visiting on vacation can now be brought to life in your living room via 3D vistas. This headset also turns your photos and videos into 3D objects. Sitting at home, you can be transported to fantasy worlds. Thanks to advances in AI, immersive experiences like this are now greatly enhanced. [26a]

Some headsets are costly, no doubt. However, it's been reported that companies such as CEEK VR offer affordable options for beautifully immersive experiences such as concerts and sporting events as well as immersive audio experiences which (per their website) is the world's first 4D Smart headphones. [26b]

In our ever expanding AI world, you can be transported into realistic worlds you might have only dreamed of visiting, and you can do so from the comforts of your couch.

Speaking of transporting things using AI, airspace will look quite different than we know today. In the near future, your ride-sharing experience might include an air taxi landing in front of your house to pick you up. The FAA has said that it will be issuing permits for their operation in as early as 2025. The FAA released a blueprint for Air Taxis in early 2023, and is moving ahead rather speedily toward this new reality. [27,28]

Want to avoid expressway traffic? No problem. Call an Air Taxi. Other countries are also headed into similar futures.

A UK company called the Vertical Aerospace Group will debut their VX4 model at the World Expo in 2025. [29] The cruise speed is 150 miles per hour, according to their website. [30]

On the ground, other automotive technologies are also in the works. In 2023 Ford Global Technology filed for a patent for AI technology that would allow a car to repo itself when someone becomes behind on their payments. [31]

According to the patent—"after warning of non-payment, the financier of the car would be able to use the technology (via the cloud) to begin shutting off car components such as the radio and air conditioning". And, if payments still are not made, the car could be instructed to drive itself back to the finance company or bank, etc., by using its autonomous driving system.

Meanwhile, cameras in the car would be used to see if the car is inside a garage, or who is around it. Talk about AI and Algorithms expanding further into our everyday lives.

On a much more positive note, if the driver of a car using the Ford technology had a medical emergency, the car could be instructed—via the cloud—to take them to a hospital. [31]

This would be a great use of AI technology. However, if the person's payments are behind when the car drops them off at the medical facility, they might want to start thinking about an alternate ride home because the car will probably have other plans such as driving itself back to the dealer or finance company.

Criminals have not been left behind on our AI journey. I heard of a police chase recently that was ended by the authorities bringing the car to a complete stop via a cloud-based service.

It has already been widely announced that the U.S. is moving toward the requirement of self-braking systems in all new cars.

Would you be more comfortable in your car knowing that it might potentially help you avoid accidents by stopping on its own? Or, would you be completely uncomfortable knowing that your car will be handling braking decisions for you, on its own, when it chooses?

With self-driving and self-braking cars, there are many concerns to be considered. Could a hacker suddenly cause your car to halt or even reroute itself?

This is serious stuff! This is why there must be serious oversight of such technology and the corporations that control it.

Will there be times when it would be critical for you NOT to stop but your car calculates that it should and slams the brakes right then and there, regardless of what you want?

Self-driving autos have had their issues. There have been cases of some cars getting confused and making tragic mistakes.

Remember, AI machines and apps can't actually think or see. They are calculating their actions and activity based on data inputs.

Self-driving taxis have already joined us as a new species of traffic jammers. Some have famously disrupted traffic. One brand of self-driving taxis (called robo-taxis) began operating in California in 2022. [32]

A self-driving car decides what to do, on its own. The car can calculate when to apply the brakes because it is always using sensors and cameras to detect the activity around the car. It is constantly calculating what it should do.

It is quite something for a machine to be able to decide its own behavior as traffic conditions and road situations constantly change.

Is it possible that a self-driving car can be pre-trained to adapt to every possible situation that it might ever encounter in the future?

So far, the record speaks for itself.

There have been times when a car's AI system failed to make the same logical decision that a human would have made had a human encountered that same circumstance while driving. There are times when self-driving vehicles have made disastrous errors such as near-misses with trains and other incidents.

As for the driverless taxis I just mentioned, I could not finish this book fast enough before things have already changed. By November of 2023, the California Department of Motor Vehicles had indefinitely reportedly suspended permits to continue operating driverless taxis for the company Cruise, owned by General Motors. This followed after one of the taxis allegedly rolled over a pedestrian. [33]

According to a statement on the DMV's website at that time, the vehicles were not safe for public operation. The DMV further stated that the "manufacturer had misrepresented information related to the autonomous technology of its vehicles".

Again, in a world where AI machines and apps are given the power to be our co-thinkers—we must keep in mind that the machines have no understanding of what is right or wrong, good or bad, decent, inappropriate, moral or whatever. AI simply carries out actions based on calculations and algorithms. Again, these machines are not sitting there pondering philosophies about the value of human life. They couldn't care less. That's still our job as humans.

Corporations are creating these AI products behind the private doors of their own companies and their products can affect our lives in critical ways. Sometimes our lives are literally "in their hands" as they create machines to, effectively, think and reason and make decisions for us.

As I said earlier: As a human group we must ask ourselves and each other, "Exactly which activities can be safely or adequately trusted to AI machines whose 'thoughts' are limited to managing data based on probabilities, calculations or statistics? What issues will we allow AI machines to think about and 'decide' for us: safety issues, medicine, science, law, or our civil liberties?"

Actually, these questions are arriving late to the party.

Long before now, the activity of machines thinking for humans slipped into society. It entered quietly through the backdoor via social media and Internet algorithms.

Algorithms have long been funneling their own versions of reality and truth to us. On the Internet, algorithms have long been working around-the-clock to shape our perceptions of ourselves, other humans, and the world around us.

Even beyond the Internet, AI algorithms, are already managing our human affairs on some of the most critical levels of society. At this point, many decisions about our Civil liberties and privileges as citizens have already been handed-off to AI Apps and algorithms to do our thinking for us.

Here's an example of where AI and algorithms have gone terribly wrong.

Imagine going through your life minding your business—working hard to take care of your family. One day, out of nowhere you are stopped and arrested because AI decided that your facial biometrics matched-up with a criminal who was wanted for committing a theft. Of course you are innocent, but an AI program has decided to choose your face to plug into a crime scene. As a result, the authorities show up and arrest you—all based upon algorithms calculating-up a truth all their own—even though they are not capable of understanding human truth at all.

This has actually happened to real law-abiding citizens who were simply going about their lives not bothering anyone.

Yet, algorithms are currently used in areas of society where arriving at the actual "truth" is the most critical to civilization. In the case of the arrest, an app was permitted to serve as a gatekeeper of the truth in an area where upholding truth is most vital and where it is most critical for humans to remain in tight control.

In our current reality, AI is making decisions in areas that will mean someone's continued freedom or the complete loss of it. It is deciding what the "truth" should be even though it has zero understanding of the concept of truth or how it affects humans. Yet, these machines and programs that were created by corporations and companies, are out there reaching conclusions for humans, every day. All while having zero accountability to humans and their laws.

AI Algorithms have long been deciding critical issues about our lives. They are deciding things such as who will – or won't get the home and auto loans; the job interviews, college acceptance letters, credit card approvals, on and on.

Is this acceptable to us as a society? Should machines be left to control these decisions?

These are vital issues we should think about as we think about the candidates we send into office to protect our interests as humans.

Many citizens have no idea how deeply embedded AI is in their lives, and that AI and algorithms are involved in many serious matters as they operate among us on a continuous basis.

An AI machine—that is neither a human nor an eyewitness to any crime—could potentially decide that you are a criminal and, from there, the law will act on the machine's decision as if the AI machine has witnessed you committing a crime. In actual truth and reality, you were not involved in any crime, still it's possible for algorithms to be given the authority to make such decisions. [34]

This has already happened to some.

In these cases, it is important to understand that data crunching AI machines—fed by a limited amount of facial feature samples—used fault-filled, pattern-matching, algorithms to pluck actual humans out of their lives where they were living as law abiding citizens, and match them to "truths" imagined up by algorithms.

How? ...

It's because machines (or programs) were allowed to replace human thinking and reasoning with algorithms that did the thinking for the humans. The algorithms used image data—on their own—to match the wrong person with a crime that the person was nowhere around while occurring. The algorithms fabricated an "untruth" that triggered actual lawmen in real life to jump on the trail of perfectly law abiding citizens and to track them down, and jail them. This happened to real people who were going about their business. [34]

This goes way beyond a machine assisting humans with their tedious tasks. This is an example of a machine being elevated to an overseer of Human Rights.

Shouldn't the business of managing Human Rights belong to humans that are thinking carefully about every aspect while they examine every detail including all extenuating factors?

And, shouldn't these humans who are responsible for overseeing our Human Rights be accountable to the laws, themselves? Algorithms have no accountability. They just work correctly or they don't. So, trusting an AI machine to decide who gets to have their Human Rights protected or who is excluded (due to error) is something we should all be deeply concerned about.

When AI machines produce inaccurate information, the errors are not always easily detectable by humans because we are already beginning to rely heavily on them without double-checking the results.

In the arrest cases just mentioned, there was certainly a point where the humans involved in the arresting of innocent people might have caught the algorithms' mistakes right away. Yet, for far too long, the mistakes continued to play out.

This proves how quickly some people can be biased by AI suggestions, believing AI algorithms to be reliable and unlikely to make mistakes.

In these cases, corporations tell customers what the algorithms are capable of. They show these customers some amazing examples of how the algorithms might help them and they close the sale. Then the customers implement the AI programs and rely upon them as if they are fault-proof.

Lawmakers rely on AI for a myriad of reasons. The result is that corporations have now become a part of the equation of dispensing and managing Human Rights and Civil Liberties, via the products they sell to law-involved institutions or other entities.

Again, corporations are creating AI algorithm-based apps that automate critical decisions about our lives and freedoms. And, some of these apps are effectively doing the thinking, reasoning, and rationalizing for humans in areas that should only be reserved for "thinking" humans. Or to be more specific, they should be reserved for humans who are thinking real. real hard about these matters. These critical-decision areas should not be handed off to machines with zero consciousness, empathy, or care for human life and activities.

To those that steadily argue that AI disrupting civilization is far-off in the future and that we don't need to be overly focused on those risks right now—I invite them to imagine the dystopian present (not future) that those falsely-arrested innocent people experienced as a result of an AI app's mistake. It was the AI app that decided their fate on those days. It was the AI app that caused actual humans to be arrested and locked in cages—one man for 5 days and nights, another for nine—one an expectant mother. And there are other such stories. What could be more dystopian than that?

Again, this is why no human can afford to not understand the subject of AI. Many very serious matters of society and civilization rely on decisions that AI apps and machines are making, regardless of whether AI is getting the answers wrong or right. And AI is now making decisions for humans on some of the most critical layers of humanity. Our justice system is only one among many other areas.

Some cities and states are now implementing protections against AI in areas such as employee recruiting—particularly during screening and hiring, and its use in the workplace. [35]

In addition to employment, AI has proven problematic in other critical areas.

In 2023, the biggest healthcare insurance provider in the U.S. was facing a class action lawsuit for allegedly using AI algorithms to wrongfully deny coverage for elderly patients requiring nursing home stays or stays in nursing facilities. [36] The error rate of the AI in this case is alleged to have been 90%. This percentage error rate number was allegedly derived from those cases that were appealed and then later approved by the insurance company. Appealed cases were approved at a rate of 90%, indicating a tremendous level of inaccuracy in the AI algorithm that initially denied those same cases.[37]

Knowing this, I hope seniors and their children will appeal denied claims. This is one example of how individuals can stand up against the detrimental effects of AI automated inequality and unfairness into our human existence. Challenge the decisions that are automated by AI if you feel they are unfair.

These kinds of issues are exactly why I say that anyone who is out there assuring the public that the critical risks of AI are still far off into the distant future is someone who is not in-touch with the present world where AI is already performing as puppet masters of human affairs in many areas. Either they are uninformed or I would suspect that they stand to gain personally from diverting the public's attention away from conversations that could lead to heavier scrutiny and regulation of AI-creating corporations.

At this moment, we might not yet have to worry about armed AI robots escaping their labs and taking over the world, but we certainly have to worry about AI and algorithms inciting humans (who are armed) to chase other humans down, arrest them, and lock them inside cages. And we certainly do need to be concerned about our healthcare, financial systems, and many other critical systems becoming robo-controlled by AI machines as they quickly become the controllers of society, thus, civilization. The unjust law and healthcare scenarios mentioned here are not futuristic "what-if" scenarios. These scenarios are the "what-is".

I'm sure that none of us object to corporations creating products that make our lives easier; but, having corporations solely in-charge of assuring the integrity of AI machines and algorithms that ultimately dispense our rights as citizens, or control our healthcare, or think for us, or distribute their versions of intelligence to us, and control what we believe is fact—this is another territory, altogether.

If we are to thrive in an AI world, we must become wise enough to quickly spot the difference between AI apps that make our lives easier, safer and healthier as opposed to those that quietly steal away our liberties and leave us as a bystander species in our own story of humanity. We don't want to end up like baby birds in a nest, robotically auto-fed by AI machines that do the thinking for humans.

Again, whether the machines are getting the answers right or wrong, many professionals are relying on them to deliver critical information and conclusions.

Some professionals are learning the hard way that AI Chat sometimes makes things up that sound very legitimate—things which turn out to be completely untrue.

Allegedly, as part of a court case, an attorney referenced prior court cases that had never taken place. [38] The attorney eventually had to admit that he'd relied on AI Chat for research. AI Chat had given the attorney what seemed to be credible results when it supplied him with fake court cases, but AI Chat had simply made the information up. Think about that: Someone's rights during a lawsuit could have been affected in court, based partially on meaningless information that a machine just "decided" to make up.

This is a big deal in a society where court cases decide the allocation of both rights and liberties.

After this incident of AI giving an attorney incorrect information, one would think that all legal professionals would be extra careful while using AI to assist them. Apparently, not all did. Another high-profile attorney was soon all over the news for allegedly submitting false reference cases to his own attorney. Those cases had also been hallucinated-up by AI. This time it was by an AI search engine.

This is, yet, another example of AI's potential impact on human intellect and knowledge.

In my opinion, AI did much more than merely provide false research to these attorneys. Instead, the AI machines ended up producing their own versions of human truth. The machines "decided" what kind of cases it should conjure up for the attorneys. Effectively, the AI machines did some of the critical thinking for these attorneys, and what it "thought up" was submitted to real courts and substituted for actual truth until other "thinking" humans discovered that the cases were fake. These attorneys were not new to their professions. They were not naive people who might have been easily led astray by AI. It's reported that one lawyer had been practicing law for 30 years. [38a]

These stories show us AI Chat's potential to lead intelligent people down a path of falsehoods. AI Chat and AI search engines are already having an effect on humanity as humans rely on them to do our thinking for us. Who knows how much false information is going undetected as machines supply information to professionals on every level.

This doesn't mean that AI Chat or other machines should never be used for research.

What it clearly means, however, is—we need to manage these machines as they assist us. We need to verify the information they give us … especially where that information is used for important reasons.

This brings us to another important question:

Do you feel your lawyer or doctor, etc., should be required to disclose when they have used AI Chat to perform some of the work for which you are paying this professional to perform?

Personally, if I hired a professional based on their reputation for brilliance, I would want to know if they planned to hand-over part of the job to a hallucination-prone AI machine. I don't want to imagine this professional racking up billable hours for what appears to be extensive research when, in reality, an AI Chat machine is performing that research for them in billable seconds.

As for AI machines and their algorithms dispensing misinformation, the warm-up act was surely brought to us by social media and Internet algorithms. Today, AI Chat machines could potentially catapult the spread of misinformation to levels so bizarre that only a cartoon characters could pull-off such a movie plot.

Corporations and their algorithms are steadily cornering the market on overseeing humans. Already they co-own and control our private data as well as shadow us everywhere we go, via their devices.

In fact, we are not even permitted to use the products we buy and own without first clicking those "I agree" buttons containing endless, complicated contract pages that even the Supreme Court would have trouble deciphering. What the contracts should say very simply is …

"If you use this app or search engine, we will probably follow you around in your life, track your online activity or copy every password you type—forever more—or sell your information to others so they can target you constantly with ads? Do you agree with this or not?"

The more I discuss the subject of AI and corporations, the more that cartoon movie poster seems to work—the one that depicts some of these major corporations as mustache-twirling mad-scientists vying with each other to take over civilization for profit.

Too many times we've seen corporations put profits ahead of what's good for humanity or side-step rules designed to protect us?

There was once a time that no one could steal your identity or life savings without having to physically steal your wallet or purse, or either rob a bank in-person. Today, people on the other side of the world can rob you, rob the bank, or steal the identities of tens of thousands of people with a few key strokes on their cell phone.

Some might say that this is where technology has brought us. But, the truth is, this is the place technology-creating corporations have brought us to.

It may seem that this state of human affairs is the inevitable byproduct of technology and it just naturally evolved out of the technological ether, over time. But it didn't. People in Boardrooms and conference rooms made decisions about data networks and Apps that created many unnecessary security vulnerabilities in networks, servers, and other digital systems. Now, it is impossible to go backward and rewrite those chapters in technology.

Despite the fact that security technology struggles to catchup, the past damage is undo-able.

In corporate meetings, there have always been those certain people who argue for the roll-out of products long before those products have been made safe for the market. The chase for profits overtakes them. The pressure to make good on earning projections overtakes them. The fear of losing that huge paycheck or bonus overtakes them. The pressure from higher-ups overtakes them.

Too often, this leads to stressed-out pressurized CEO's and execs who release products with known problems built in—products which end up creating unbelievable harm to the public. Yet, the problems were ignored or hidden by execs who conveniently "closed their eyes to keep from seeing" them. Meanwhile, the public is stuck with the fallout.

This is one of the reasons why the word "recall" is now part of the everyday language of commerce.

Corporations have constantly been required to retrofit technologies with better security after they have hastily rolled out products without giving adequate thought to limiting harm to users. These days, they seem more comfortable with announcing that they let some hacker walk away with our personal identities, than they are with assuring that they release products on platforms where this cannot happen.

However, because of AI's immense and unprecedented power, it will be impossible to stop the damage once some AI products hit the mainstream. AI will neither leave room for a landscape of corporate neglect nor greed.

I must also say that it's very easy to throw around the term "greedy corporations". The fact is, for-profit corporations are made up of scores of shareholders; "humans" who invest in them hoping for their stock shares to gain more and more value. There is nothing wrong with profiting from an investment, but shareholders will have to hold corporations more accountable when it comes to AI development.

If we are not careful about how we allow corporations to implement AI throughout our lives on every level, it could become too powerful, too fast with unstoppable or irreversible effects.

This may sound like a stretch of the imagination, but great scientific minds are very concerned about just such issues. Surprised by the potential power of their own creations, many AI developers, scientists, and corporate CEO's now openly acknowledge the threat of certain AI technology possibly ending up in the wrong hands. Some corporations are now appealing to Congress for AI to be regulated.

When I first began writing this book, many people—including some in Congress—were predicting that AI would remain largely unregulated. To me, this did not seem possible, given its power and potential.

Today, things have changed.

Recently, one AI Chat machine creator has posted instructions on its website showing how to block their machine from using your website content to further train its Chat machine. Such actions may not un-train the machine from utilizing all of the knowledge it has already learned from Internet content, but it's a start in acknowledging that humans have a right to be protected from corporations slurping up their content for its own purposes.

Thankfully, some attempts are now in progress to minimize the risks of AI development and use.

We certainly cannot afford to enter a future of advanced AI where corporations and regulators are scrambling to do damage control after the fact the way they are doing with social media and Internet algorithms; or trying to correct security deficiencies after it's too late to reverse the damage.

As we move into a future of AI machines, adaptations will be required on everyone's part. For example, we will all have to become proficient at the input language of machines or "Prompt" writing.

(Prompts are the words we use to talk to AI machines and to convey our questions and requests to them.)

For example, we might type a Prompt to ask a machine, "What is the most used word throughout the globe?"

If you change one word in your Prompt, you could possibly get a totally different response from the machine. The better you are at figuring out what Prompts the machine needs from you in order to give you the exact information you seek—the better results you will get from using AI Chat machines.

If your Prompt requests an AI Chat machine to, "Name the most important property of ice"—the AI Chat machine might answer, "Ice is cold" or "Ice is slippery". But, if what you really need to know is what properties make-up ice, then the machine would have just given you useless information due to the way you wrote the Prompt.

This seems a simple thing to detect, but when we are researching subjects we know nothing about, we may not recognize when the information supplied by a machine is useless. Learning to speak to machines in their language will better enable us to get what we need from them.

In many ways, it is not the AI machines that are like newborns joining us Humans, but rather, humans are the newborns that are joining the AI machines' world. Just as toddlers observe adults in order to learn how to get what they need from them, humans will need to observe the feedback we get from machines in order to learn how to get what we want from them.

As AI machines expand quickly into our lives, efficient Prompt-writing will be a form of literacy. After all, we will need to know how to communicate efficiently with our new companions.

Thinking of machines as our companions may have once seemed odd, but corporations have rushed to facilitate this shift by giving their AI machines human-sounding names. This helps feed into the tendency of many people to confer human-like qualities to inanimate objects.

Thanks to the electronic toy industry, some children are being exposed to AI as early as infancy. Playful AI robots have no problem smiling blankly while a child rambles on endlessly about their favorite cartoons.

AI robots gain the affection and confidence of children in no time. They become "best friends" and companions. After all, what child wouldn't think their robot "friend" is not real—or part human, at the very least?

The feeling that they are interacting with something that's part-human is not limited to children. Many adults will also form these connections.

While engaging with some AI Chat machines, it can be hard not to slip into the sense that we are engaged in conversation with something that actually cares about the conversation.

If you've ever gone online to return something or to complain, you have probably already treated a machine as if it were human. When the chat assistant replies with, "My name is Bob. I'm sorry you had trouble with our product and can imagine your frustration. I will correct the problem right away." If you typed "Thank you", you likely just treated a chat-bot like a human being. And whether or not you knew it was a bot, that chat window will close if you don't type a response ever so often. Besides, these customer service bots only interact with you in human language, so you have no choice but to talk to them like they're part of the human family. If not, they'll have the nerve to cut you off.

So, we are all out here talking to bots on the Internet and wishing them a "Great Day". The last time I had to return something Online for a refund, the chat-bot told me I was the most patient customer it had talked to all day. How could I not type, "Thank you".

I'm guessing our propensity for treating machines as though they are human started when automated voices on our telephones first began instructing us to, "Press '1' now". And we politely complied. However, the automated voices kept repeating the command until finally we yelled, "I did press '1' you idiot!" And there, began our dysfunctional love affair between humans and machines.

Now, we have graduated to yelling at our GPS systems when they keep insisting we make a U-turn when no one will let us over to make it. And we have out and out tantrums on our computers when they are too slow or freeze.

As humans, we have a propensity for re-imagining inanimate and animate objects into real entities. It's why our eyes water at the end of the "E.T." movie.

Yes, we know—on a logical level—that "E.T." is not really a space creature who dropped by Earth to audition for a part in a Steven Spielberg production. And when Spielberg yelled "Cut!", E.T. didn't waddle off to his trailer to read flying saucer comic books while they setup the next scene. But, the fact that E.T. mimicked familiar, human-like endearing qualities was enough for audiences to form attachments to the machine as if it were a real actor.

When it comes to AI Chat machines letting us drone on endlessly about every random thought, interest, and question we ever had, these machines can easily start to feel like devoted assistants or companions.

The backlash from some in the public was immediate after one AI Chat machine was adjusted to make it less prone to using offensive language while replying to users. Some users hit the ceiling. Many flocked to social media to pour out their emotions, behaving as if their close friend had been gagged, silenced, and dragged away by its corporate creator. User's reactions ranged from angry and resentful to disappointed, and some people sounded as if they were experiencing the end of a relationship.

In many ways, these kinds of responses are predictors of what's to come with AI's expansion. Our emotional connections to machines will deepen, even beyond the intense connections we already have with our cellphones. Clearly, most of us can't imagine spending a day without them.

AI apps, in their many shapes and forms, will continue to expand in their roles as our assistants whether in our homes, cars, or on our jobs.

Some experts agree that AI will affect jobs in every industry. Other experts disagree.

Some insist that—rather than replace workers—AI will become more like a tool to assist employees. However, if you are given a new AI app that acts as your assistant and this now means the chores of your 8-hour work day can be done in 4 hours, do you think your boss is going to say, "Feel free to take a half day off with pay from now on." Instead, you will inherit more work to do. And the more work each employee can do utilizing their new AI pal, the fewer workers are needed. So, I don't think we can afford to ignore the potential of AI to disrupt employment in several areas.

Already, the UK is looking to use AI to reduce the administrative work of teachers and police officers by using AI to perform administrative duties.

Plans to invest, or continue investing in AI are a top priority for most corporations, worldwide. Money market reports lead with the subject of AI and this topic is at the top of discussions among world leaders during every global summit.

Developments in AI will change things in ways no one can predict. But, as mentioned earlier, we don't have to look into the future to understand the potential of algorithms to affect humanity.

Thanks to social media sites, Algorithms are now co-parenting our youth. And, for some time, they have been influencing how Humans think about themselves and the world around them.

In a mere generation, "We The Humans" have been sub-divided and re-classified by corporations running Algorithms. Algorithms are deciding our stations in life online, and that spills into our physical world.

Search engines, social media sites and other Apps, have long been using AI and algorithms to curate our lives online. Effectively, they have long been deciding "who" will be exposed to "what" content. Again, we have gradually transformed from a society of "We the People" to a society of "We the Algorithms" and advances in AI only turn-up the stakes.

If you think algorithms are doing a good job of influencing Human thinking, just wait until AI Chat systems and other Generative AI tools really take hold of the mainstream.

For this reason, Algorithms deserve a conversation all their own. So, let's briefly discuss them.

Chapter 4

(Algorithms: The Real Social Media Influencers)

When it comes to AI Algorithms helping children to sort out what's real or not, or true or not, on the Internet—human guidance often has the equivalent impact of an umbrella in a hurricane.

Social media is the Sun around which their worlds revolve.

In fact, when it comes to AI and Algorithms, you can think of Humans as the "Followers" and AI Algorithms as the "Influencers".

In addition to AI Chat dispensing its own version of the truth to us—at lightening speeds, there are some search engines now employing AI Chat type interactions. They are also distributing their own curated versions of reality to us.

AI search engines don't just receive your search terms and respond with website results. Now they pretend to be human just like you. They appear to be thinking, reasoning, and guiding you toward answers—like a trusted friend who can accompany you like a pal as you sit on the Internet seeking answers to your questions or looking for products to meet your needs.

Today's AI search engines are right there, guiding what information enters our brains and those of our children.

Algorithms everywhere online have been stealthily working for some time to reduce human existence to one nonstop marketing blitz. To them our existence as a species seems to represent one single purpose ... one possibility—more eyeballs on ads.

Algorithms have gradually distilled humankind down to one major group—the "consumers".

Social media has then further sub-divided us into two supporting categories—the "followers" and the "influencers".

Today, new generations are born into a norm of wearing labels such as "followers", and society has gradually fallen in line and accepted this re-categorization of humans by corporations.

Corporations seem almost cartoon-level diabolical as they reduce human existence to one long commercial filled with jingles and catchy phrases. All the while these commercials—pushed steadily to the forefront by algorithms—are too often designed to make us feel as though we are failing to be suitable humans unless we purchase certain trendy products, services, or body enhancements.

Nonstop, algorithms are working to keep our youth attached to their cellphones with umbilical-cord-level dependence. Meanwhile, they supply corporations with our most private data so corporations can better target, track, and study us. Algorithms have helped corporations to study us very well. They know how to keep us swiping and clicking. They know on a scientific and psychological level how to keep us engaged.

Lawsuits have been brought against many major corporations based on the alleged addictive nature of social media Apps.

The longer you are on websites clicking links or "like" buttons; or commenting, reading, watching content, and talking to your friends—the more info is learned about you and the more data algorithms have to work with to analyze you and predict the products you might want or need. This makes you a better target for ads and products. Our search histories and buying habits makes us more promising targets for the corporate overlords of our online existence where achieving Return-on-Investment (ROI) is the only real law of the land.

As algorithms learn about us and direct the flow of information it chooses us to receive, they can limit our experiences on the Internet by type-casting us into categories. They choose where we fit in or don't and they tailor our online realities for us. Many of those online realities morph into our "real" lives. The confidence of a whole generation now depends on how many people "liked" their photo or post, or who "followed" or "unfollowed" them. Mean comments on social media pick at their self-esteems or leave them feeling like outcasts.

Teen girls looking for diet information online are whooshed away to virtual lands where they are fed images celebrating anorexia. Social circles comprised of teens with severe eating disorders who are shown celebrating their conditions, appear in teen news feeds. Teens are targeted by non-stop videos of how to avoid facial wrinkles. What teen has facial wrinkles? The constant mental conditioning of "you are not good enough, but here's the product we offer to save you" is relentless.

The great news is:

Parents are now rising up and bringing lawsuits against social media platforms.

Today's youth live in a generation where algorithms efficiently sort and categorize their value based on social-media attention and the number of followers they gain. Meanwhile, idiotic or over dramatic behavior is rewarded by algorithms with higher rankings that cause them to receive even more social media exposure. Young people confuse this added attention with popularity and adoration.

It is important for all of us to understand this one thing very clearly: Algorithms are not just some cute little processes that help us shop via product suggestions. They are active participants in our lives. At lightening-speed they are creating and shaping public opinion through automated manipulation. In doing so, some algorithms are manufacturing new perceptions of reality out of thin air. They are creating new beliefs around what humans think of as significant or worthy of attention.

In short, some algorithms are helping us to form new and skewed perceptions of reality that would not exist without their nonstop input.

In the Senate hearings on algorithms [39] executives from Facebook, Twitter, and Google attempted to praise their companies on the so-called great strides they have achieved in making algorithms safer for society. In addressing the extreme misuse and abuse of algorithms by the aforementioned corporations (whose execs were busily testifying what a great job they were doing) Tristan Harris, the Executive Director and Co-founder of the Center for Humane Technology [40] said the following:

"A business model that preys on human attention means that we are worth more as human beings (and citizens of this country) when we are addicted, outraged, polarized, narcissistic, and dis-informed because that means that these companies have been successful at steering our attention using automation." [39]

BAM, CLANG, POW! Insert mic drop!

We need more moments like this in Congress where corporations are required to face public and Congressional scrutiny concerning how their technology is affecting us as a civilization; especially as technology-based products become implemented on the level of affecting our Civil Liberties as well as our access or control of information online.

During the Senate Hearing, executives took turns defending their corporations in tones so warm and kind that you might have imagined actual algorithms processing inside their brains to keep them non-combative and laser focused on winning the heart, thus approval, of the Senate. Each one stuck meticulously and robotic-ally to their marketing message.

Listening to them testify—some might have even been convinced that these corporations were only collecting our private data in order to take care of us because they care so much about us.

They care so much about our precious time that they want only to help us by "personalizing" our online experiences.

But, the truth is, there's a big difference between a corporation using algorithms to make friendly suggestions to us about what to read—as opposed to corporations quietly detaching the whole of society from free-choice in our online searches for information or products because algorithms reduce or exclude those choices while also eliminating our rights to privacy in both our digital and physical worlds.

Not to mention the fact that people who are harmed through social media targeting them with "extreme content" may not detect any harm is occurring at all. They may not detect that algorithms are manipulating them through targeted content, and that the content is constantly reinforcing their negative views and, therefore, serving to narrow and compress their thinking, perceptions, and views. If life is about learning and growing, then algorithms have certainly disrupted our development as a human family.

Facebook's content-ranking algorithms curate the posts on your page-feed in a way that keeps you engaged and interested. Content that stirs the pot is fed to us in ways that keeps us commenting on the content; that keeps us agreeing, disagreeing, sharing and "liking" posts.

In short, Algorithms are expert at keeping the haters hating.

In a scenario where the potential power to affect civilization is already concentrated in the hands of a few major corporations, along comes major advancements in AI; and some of these same corporations are jockeying to out-compete each other to make AI machines and apps.

AI (unlike any other technological framework in history) will force humans to confront the fact that—what happens in our digital world is predominantly what controls our physical world.

Our digital planet is no longer just an extraneous place we get to play on as we surf the Internet for our interests. Instead, our digital Planet Earth is more like the star around which our physical Earth revolves. And that star is covered with corporate logos.

As we've evolved into a global digital society inter-connected by data, corporations have moved stealthily to siphon-off our rights to privacy and to overlord the flow of information. Too often Algorithms have been implemented at the expense of our basic rights.

Imagine if you heard about Congress introducing a bill that would require all Americans to check-in with the government anytime we wanted to log onto the Internet, or check emails, or log onto our bank accounts, credit card companies or make an online purchase. Imagine that the bill further stated that the government would accompany us during every website visit and keep records of our passwords and every single thing we did while logged on. Finally, the bill would state that we have lost our constitutional right to be free from unwarranted search and seizure and that the government would track and record our locations at all times including keeping track of our every digital interaction while in our cars, or using our cellphone, computers, or any electronic device.

If ever such a bill were to be introduced in Congress, most of us would rise-up immediately.

Yet, all of this same critical power has been gradually acquired by corporations using algorithms.

This monopolization of our identities and privacy by corporations didn't require one court order. It was just done to us, one App at a time.

Our digital planet spins on an axis spun by the index fingers of a handful of major corporations. Advances in AI have highlighted just how far humans have permitted corporations to go using algorithms.

Meanwhile, advances in AI serve as a point of demarcation where humans have been put on notice. It confronts us point-blank with one choice: control machines or become controlled by them. There is no third option.

Hopefully, humans are not willing to give up any additional personal liberties to corporations, let alone new AI machines. As it is, our virtual footprints are nonstop tracked and traced by corporations; and our personal data is constantly collected and combed through by them.

The term "personal privacy" is an old saying these days. Ironically, the ones who use this term most often are the very corporations whose "I agree" policies have demolished the concept of privacy.

We don't even let those closest to us in on our bank passwords, yet we tolerate giant corporate strangers watching over our shoulders during our most personal online transactions?

The reason we've allowed them to do so is simple: corporations have been permitted to coerce us into signing away these basic rights to privacy in order to use the Apps and devices we purchase from them. It is absurd that we can pay for a product, yet not be allowed to activate the product unless we agree to let the corporations have unlimited permission to invade our privacy forever more. That is equivalent to buying a car, but, just before the dealer hands us the keys, they demand to know everywhere we will ever go in the car, how long we will stay there, who we will interact with while there, and any stores we might visit along the way, including a list of what we purchase. Actually, with today's cars, the dealer doesn't have to demand this information. The cars are busily reporting all kinds of information to App makers and manufactures, on their own.

Meanwhile, corporations are sending nonstop announcements to assure us of how wonderfully they are protecting our privacy. Yet, these announcements come at the end of their hacker alert announcements notifying us that they have just let some hacker skip away with our most personal and critical data. On top of that, when you call the company with your concerns, AI chat bots are happy to greet you and to pretend to imagine your frustration and how you feel (as a human) to have your identity stolen.

There are now generations that have never known what it is like to completely own and control their own personal information. This used to be one of our greatest rights and privileges, yet it has been slowly confiscated by corporations. With advances in AI, it is now more important than ever for every generation to understand what's at stake as technology evolves. We must question what other rights might be gradually siphoned away from us by corporations as we happily click and swipe and agree. Owning products should not require us to give up our basic rights and privileges. There should be limits to what a corporation can require us to agree to in order to use their Apps.

Again, if humans don't join the AI conversation and become more vocal about their concerns or desires, AI could gradually relegate us to bystanders in our own story of humanity.

This is why the question of knowledge versus intelligence is more critical than ever.

Given that both the real planet Earth and our digital planet Earth are controlled by data, what does it mean if major corporations and their AI machines come to control all the ways in which humans can access information online?

This is a major question that should be raised and discussed from exactly this perspective. Especially as AI Chat machines and search engines expand to help spoon-feed us everything from content to news.

Currently, there is a major media outlet suing a major AI corporation for using the outlet's content to train their AI machines. Words written by journalists doing the hard work of investigating and researching—while also adding their lived perspectives and consciousness to how they approach their subject-matters—are now having their articles and reports remixed by AI machines. This restructures the way in which we receive news and moves it from a human to human broadcast to a machine to human broadcast.

In a different lawsuit, a major search engine company is being sued for directing online traffic to its own product platforms, effectively cutting out its competitors as users search the Internet for products.

In our homes, many of our electronic devices can listen to our most privately-uttered words. Some TV remotes now contain microphones through which to interact with us and our smart appliances can monitor our personal habits. Meanwhile, some of our devices can monitor our conversations and even join in to make suggestions.

Many of our automobiles are collecting private data on us through onboard cameras and microphones, as well as Bluetooth interactions with our cellphones. Information—including what you do inside your car and where you go—can be captured by certain major auto manufacturers and the information can be sold to others. [41] As we use our cars, everything from call logs and passwords can be left behind in your car. [42]

While simply driving along, many of us have no idea of the data that is captured and uploaded via data networks. All kinds of information can potentially be captured including videos of what's going on inside our cars as well as images of our passenger's activities, can be captured with some newer models. [43]

Not to mention that any digital device that connects to a network can potentially be hacked.

So, Online, we have corporations and their algorithms monitoring our activity and behavior while social media algorithms redefine and recreate the human world, in their own algorithmic images.

In many ways, algorithms have long been mentally looping us into stagnation as a society? No one enjoys being wrong, still, it's in our best interest when we learn and grow from diverse points of view. Personally speaking, I want to grow from my mistakes, not have my lack of understanding or ignorance reinforced because algorithms preclude me from discovering info that helps me to see a bigger picture. I certainly want our youth to learn, grow, and develop critical thinking skills from considering various points of view.

As Algorithms are busily helping to shape what we believe the rest of the world thinks of us – or what we believe about ourselves—we will respond to some of these optical illusions as if they are real. You may be arguing with someone online about their ridiculous opinion when, in fact, you might be arguing with someone who is just having a good time trying to ruin your day because they are miserable or even just toying with your emotions for sport because they have nothing to do. Or you could be arguing with a bot or with someone from a hostile government who has been assigned the daily job of helping to stir-up chaos in your country for geopolitical reasons.

Just one social media account under the control of a bot can spread misinformation and rumors far and wide to incite anger and chaos on social media or the Internet. Some social media bot-accounts have huge numbers of followers. People believe they are following an actual person when, instead, they are following a machine whose job it is to create chaos and controversy.

When we consider how many people depend upon online sources for their news—and then you also consider the widespread use of bots spreading misinformation or hysteria—it is clear that the manipulation of people on social media, via bots, is a huge problem. In so many cases, we are out here arguing with machines believing them to be real people who are making ridiculous comments. As we read or respond to them we are becoming angrier and angrier believing the rest of the world is heartless or ignorant.

Indeed, social media and Internet algorithms affect us, our views of ourselves and others, and our impressions of how the world is operating around us. Our brains are now up against digital machines that are targeting us nonstop with info curated to grab our attention with wild behavior, shocking words and images, and sensational phrases to make us click on websites. Meanwhile, other machines make critical decisions about our lives through the automated parceling-out of our rights and privileges; or the automated denial of our rights and privileges.

Personally, I still haven't gotten over seeing an AI tech document that actually contained the term "The Humans". I can't help but wonder if the document was generated by an AI machine.

At any rate, I assume the phrase "The Humans" still means us.

I don't know if phrases like this will begin to show up more in tech documents as machines become our cohabitants (if not our sovereign leaders) but I'll be keeping my eyes on this, for sure.

Speaking of eyes ...

There are just a couple more technologies I think are worth mentioning, before I end this book.

Today, some websites can monitor our eyeball movements while using our computers or devices. There is software that allows infrared light from your computer to be used to track where your eyes are looking on the screen. This eye-tracking technology allows you to use your eyeballs to control your computer screen and keyboard just by staring at the choices you want to make. [44]

This is a fantastic option for those who—for whatever reason—cannot use their hands. Think of all the people this opens up the world to. Still, these technologies are being explored for every possible use.

Using the camera on your computer or cellphone, certain websites can study your eyeball activity as you browse their sites. [45] They can determine exactly where on the computer screen your eyes are focused as you visit their web pages.

Imagine AI being used by a website to determine what images on the websites caught your eye first, or what objects on the web page caused you to stare the longest. This information can be used to create more info about you to feed more algorithms. Imagine how your eyes might linger on an image of your favorite dish on a restaurant's website.

People often refer to the eyes as the windows to the soul, but AI can make them the windows to our buying patterns.

Again, advances in AI have brought us to a new reality. Some of the ways in which it will serve us are incredible, but there must be new laws enacted and changes to existing laws where technology is concerned.

The good news is, there are a surprising number of organizations and advocates who are out there working on behalf of us humans every day—standing up to corporations, appearing before Congress and advocating for immediate action and regulations. On the highest levels of government, people are now moving much more quickly to try to get in front of the risks presented by AI.

What remains to be done immediately is for everyday humans to get onboard with this critical conversation. I hope this book has made you feel much more knowledgeable about this topic and I hope you will be inspired to advocate or support other AI advocates in some way. Even helping to share AI information is a contribution to humanity. We are truly at a point where we, as humans, truly need each other as AI expands into every crevice of our lives.

In fact, AI has now entered the most surprising areas of our private spaces—the seclusion of our bathrooms. Yes, "smart toilets" have been developed.

For example, a smart toilet device called U-Scan was showcased at the Consumer Electronics Show 2023. [46]

This particular smart device sits in a toilet and collects information during tinkling activities to help monitor health and also offer advice for improvements. [47] It's like having a health lab in your toilet to check for imbalances in the body.

The advantage of each of us using this technology to gain immediate insight about our own health and biological processing—or to use technology to give us immediate reports about how the things we eat are affecting us, could be phenomenal. It's easy to imagine the immense benefits of early detection of health issues that will reduce doctors' visits, etc.

Smart toilet technology development is serious business. Pun intended. Smart toilet development is also the focus of researchers, scientists, and scholars at respected universities including Stanford University [48] and Duke University. [49]

Again, these are exactly the kinds of brilliant ideas we need technology to offer us but corporations must be held to the highest standards to avoid compromising our privacy and security in new and unimaginable ways.

Here, again, we also have to consider the potential problems that might occur if such technology is rolled-out without extreme security having been built into it?

For any device that can potentially connect to a data network, the question must be asked: How secure will the device be from hacking?

This is where movie ideas immediately crowd my imagination. Suppose someone's toilet data gets hacked? … a powerful CEO or Judge hiding an addiction, perhaps? ... plots of extortion or blackmail?

With such technology as Smart Toilets, data security must be treated with the utmost consideration. After all, data going between your toilet and your cellphone or transmitted via cellphone networks or computers, could potentially be hacked. Just imagine the level of violation you'd feel if someone hacked into the most private personal info you possessed as a human—your toilet data. I'm sure you'd be pretty pi….— um, never mind.

I can already predict that, somewhere down the line, our toilets will be trying to up-sell us on products or other suggested remedies, while they have us as captive audiences. It's not hard to imagine commercials coming from your toilet seat: "You're data indicates you're a little low on iron. Say yes to order XYZ now!"

I rang someone's doorbell the other day, and a commercial started playing on the speaker. A few short years ago, it might have seemed unthinkable that someone could potentially hack your doorbell camera to get onto your home-computer network and computer devices; [50] or hack into your Wi-Fi network through your smart refrigerator, or other smart devices.

But, here we are, today, living in that reality.

Hopefully, by now, you understand why this book was so critical for someone to write and why I stated upfront that it is one of the most important books you could read about AI because of the deep dive it provides into this subject. It is important for everyone to get a full understanding of AI as quickly as possible so we can insist that it serve us rather than dominate us by ushering us into subservience via "I agree" buttons.

By the way, I had also requested an AI Chat machine to create a rap about the negatives of AI on civilization. I figured—AI may not have street cred in Shakespeare's or Tupac's worlds, but it ought to have figured out something about itself by now. So, I asked the AI Chat machine to spit a rap about AI's power. It did so in less than 2 seconds.

The last line bragged about its power to "...create or seal our fate."

"Our?" I replied. You speak as if you are one of us Humans. Do you think you are human, too?"

That's when the machine seemed to lawyer-up. It quickly denied it was human and began covering its backside as if it were a corporate exec facing a Senate Hearing. By the way, this was the same exact machine that once had to be trained to stop insisting it was alive.

Alive and sentient or not, the machine pretty much echoed everything "I, The Human" have covered in this book. "AI has the power to create or seal our fate."

Corporations may insist that AI machines are not alive, but one thing they are is surely this: machines that could potentially drop-kick us into infinity if not merged into our world correctly. And that ought to be alive and sentient enough for anyone.

I will close this book with wise words from an elder because I just love the way they so easily sum up entire dissertations with a wise sentence or two.

In referring to us as a society, Professor Latimore-Peace said:

"Once we were the ones who created technology. Now technology is creating us." [51]

The structure of society is no longer just influenced by technology, but rather, it is the actual by-product of it. People used to say that "Life imitates art". Today, Algorithms imitate art and life—while attempting to control humans in the process.

Professor Latimore-Peace has cleverly summed up the critical question: What do we want technology to make us into?

This is the question each of us will have to answer on our own as well as collectively as a society. That is why I wrote this book ... to add some perspective to this topic that, in my view, was sorely absent from it.

It is not the threat of AI machines forcefully taking over the planet that we have to fear. Instead, the far greater potential threat to civilization will come from seemingly innocent Apps that we willingly allow to replace our ability to think for ourselves or the many harms that will result from us clicking those "I Agree" buttons which have already permitted corporations to reach extortion-level control over our rights to privacy.. What we have to fear is technology corporations doing what they please with our lives as they expand their power to automate and dispense our Human Rights via new AI products.

This is why this book is so critical to the discussion of AI. It shifts the conversation from a very generalized one about AI taking over the world to a topic of corporations taking over Humans... taking over our Human Rights, our Civil Liberties, our children's minds, our rights to privacy—all while bundling their corporate control over our access to critical services.

Corporations have taken over our lives simply because nothing has stopped them from doing so; not the laws or the courts. Lack of regulations allow them to run amok with our rights—largely without having to answer to any of us or any court. A handful of major technology corporations have managed to siphon-away our most cherished rights without having to get a single comma amended in the U.S. Constitution. Corporations simply gave themselves this power through technology products and applications..

This is why "We The Humans" have to fervently remind ourselves that AI is not a story of AI monsters barreling toward us with unstoppable force. Rather, AI is the story of corporations and products. It is the story of what us humans are willing to do to ensure our survival in a world of corporate-controlled machines.

As a species, focusing our attention on the safe expansion of Artificial Intelligence must be a priority. However, there is one other AI acronym that needs to be at the top of our list of urgent concerns: It is **A**ccountable **I**ndustries. This term needs to dominate all discussions of Artificial Intelligence, immediately. How will corporations be kept in check?... not just AI, but the AI product creators? That is the foremost question to answer.

As humans, we are still powerful! We can all play a part when it comes to ensuring the responsible development and expansion of AI. We can assert our voices, our votes, and our spending power. We can create or sign petitions. We can support organizations that are doing the work to speak up to our legislators. We can challenge algorithms that auto-deny our legitimate claims such as with healthcare providers. Potential or current shareholders in companies can require accountability or not invest in companies that create products that harm humanity. And, YOUNG PEOPLE... You who have mastered the game of Algorithms to the point that you can make even a house fly go viral—you are the ones with the platforms to publicly challenge corporations the most. You can start these conversations out loud and be heard. At the very least, each one of us can share information to help inform and prepare others for the expansion of AI into every part of our lives.

These are just some of the simpler steps we can take to begin to flip this Sci-Fi-level script that has become our real life. It's not too late for us to become the heroes in our own story of humanity. We owe this to each other and to future generations.

On that note—Thank YOU for reading this book! I hope it has provoked some additional thoughts on AI and left you feeling better informed on this vast framework of technology. And I hope you will recommend this book far and wide—especially on social media. Why not make those algorithms serve us for a change by making them help We The Humans to spread critically needed information as we work to save ourselves from becoming "We The Algorithms". These machines might be powerful and might even claim to be one of us now and then, but they have no idea of the powers they're up against when "real" humans choose to join forces for our own good. We are the real marvels of this planet, and now is the time to step up and seal our own fate.

Signed sincerely,

A Fellow Human

(BTW, more details about this Human can be found at the end of the book)

Disclaimer: Products and Apps mentioned in this book serve only as examples of AI technology and are not intended to endorse any products or Apps whatsoever.

Acknowledgments

To my phenomenal PMTILSG-GLP (edit-suggestions extraordinaire), and to BabyLove-CDJ, BabySweets-JRH, B-Luv- BGH, S-I-Luv-LJ, Lovely ones PJC and ANB & MB, GMC+AC; and my SweetBabies - MEJ, KLJ, JML and NAJ, NLJ & NJ...

Plus my real-life superhero forever – *DJC for your contribution to my "un-artificial" intelligence.*

Thank you all for the love, encouragement, honesty, re-reads, ideas, and home-cooked meals; and for continuing to be the net beneath me while I do this work. Most of all, Thank You for sculpting me into someone whose work is about pouring into humanity in ways that I hope will touch the minds of others in positive and meaningful ways.

Thank You wise ones-JRB, VHB, and A-one/day-ones BAB, RJC for inspiration always!

Each of you play such a special part in my journey and I look up to you all in a special way.

To the new-gen bosses SLP, BCP, VMB, NAJ, MEK, KLJ, JRH and JKC! You will do amazing things in the world. I'm so proud of you all!

Shout-out to DB (aka 25) & TB; RC, MC, QB & family, and RH & VEH

Finally ... THANK YOU!!!

To each and every reader.: Thank You in advance for helping this book to impact humanity!

references

(1) Tiku, N. (June 11, 2022) The google engineer who thinks the company's AI has come to life. Washington Post https://www.washingtonpost.com/technology/2022/06/11/google-AI-lamda-blake-lemoine/

(2) Millson, A. (June 13, 2022) Five things google's ai bot wrote that convinced engineer it was sentient. Bloomberg Technology News https://www.bloomberg.com/news/articles/2022-06-13/five-things-google's-AI-bot-wrote-that-convinced-engineer-it-was-sentient

(3) Microsoft, (n.d) Chat gpt vs. microsoft copilot: what's the difference https://support.microsoft.com/en-us/topic/chatgpt-vs-microsoft-copilot-what-is-the-difference-8fdec864-72b1-46e1-afcb-8c12280d712

(3a) Mitchell, M. (n.d.) Quote used with permission from Dr. Margaret Mitchell, Ph.D., AI Scientist and Researcher, https://www.m-mitchell.com

(3b) Sarkar, S. (n.d.) AI Industry analysis: 50 most visited AI tools and their 24B+ traffic behavior Writebuddy https://www.writebuddy.ai/blog/ai-industry-analysis

(4) Campbell, P. (December 07, 2020) New advances in neural interfaces research at john Hopkins. John Hopkins University https://www.jhuapl.edu/news/news-releases/221207-apl-advances-neural-interfaces-research

(5) Moncrieffe, M. V (Spring 2017) Apl's newest prosthetic invention lets users reclaim their sense of touch. John Hopkins University https://hub.jhu.edu/magazine/2017/spring/prosthetic-arm-with-sense-of-touch/

(6) Metalert formerly GTX Pod Group (July 7, 2022) The innovative gps smartsole from gtx: mission-critical healthcare powered by eno one. A Giesecke+Devrient Company. https://www.podgroup.com/gtx-mission-critical-healthcare-powered-by-eno-one/

(7) Envision (n.d) AI-Powered Smart Glasses Product Example www.envision.com/glasses

(8) XRAI Glass (n.d.) Life subtitled. https://m.youtube.com/watch?time_continue=6&v=iiINO368vQU&embeds_referring_origin=https%3A%2F%2Fxrai.glass&source_ve_path=NzY3NTg&feature=emb_yt_watermark

(9) Sonantic Youtube Channel (n.d) Hear val kilmer's AI voice – sonantic https://m.youtube.com/watch?v=OSMue60Gg6s

(10) Descript AI (n.d.) AI voice generator tech product example. https://www.descript.com/

(11) Thompson, A. D. (Jan 2023) First look – elevenlabs – text-to-speech like amazon polly, sonantic.lo, AI voice cloning. https://beta.elevenlabs.io/ see also https://youtu.be/19ByKqXB8db

(11a) Johnson, D. (Aug 10. 2022) What is deepfake? everything you need to know about the AI-powered fake media. Business Insider, Insider Reviews https://www.businessinsider.com/guides/tech/what-is-deepfake

(12) Lamarre, C. (September 6, 2023) A.I. generated song that mimics drake & the weeknd submitted for grammy consideration. Billboard https://www.billboard.com/music/awards/ai-generated-drake-weeknd-song-grammy-submitted-1234507148/

(13) Youtube.com (n.d.) search term Presidents ai voices https//www.youtube.com

(14) Better Business Bureau (September 15, 2019) BBB warning: AI software now adds new wrinkle to email phishing scams. https://www.bbb.org/article/news-releases/20629-bbb-warning-ai-voice-software-adds-new-wrinkle-to-email-phishing-scams)

(15) ABC News GMA Team (April 13, 2023) Mom warns of hoax using AI clone. https://www.abcnews.go.com/GMA/Family/mom-warns-hoax-ai-clone-daughters-voice/story?id=98551351)

(16) United States Congress: 118th Congress - House Bill (2023-2024) H.R.5586 – deepfakes accountability act. (https://www.congress.gov/bill/118th-congress/house-bill/5586/text)

(17) New York City (October 2023) NYC AI action plan. https//www.nyc.gov/content/oti/pages/reports

(18) United States White House Briefing Room (October 30, 2023) Executive order on the safe and secure and trustworthy use of artificial intelligence. https://www.whitehouse.gov)

(19) OpenAI (n.d.) General product example and description AI image generator, dall-e-2. https://www.openai.com/product/dall-e-2

(20) Stable Diffusion (2023) Class action lawsuit pdf. https://stablediffussionlitigation.com/pdf/00201/1-1-stable-diffusion-complaint.pdf

(21) Getty Images Newsroom (January 17th, 2023) Lawsuit against stable AI statement. https://newsroom.gettyimages.com/en/getty-images/getty=images-statement

(22) Electronic Privacy Information Center (January 9, 2023) Ninth Circuit Revives Children's Privacy Class Action Against Google. https://www.epic.org

(23) Grieshaber, K. (June 10, 2023) Can a chatbot preach a good sermon? hundreds attend church service generated by ChatGPT to find out. AP News https://apnews.com/article/germany-church-protestants-chatgpt-ai-sermon-651f21c24cfb47e3122e987a7263d348

(24) The Art Story (n.d.) Fernando Botera Columbian painter and sculptor. https://www.theartstory.org/artist/botero-fernando/

(25) Aiva AI Music Generator (n.d.) "General product example and description" www.https.aiva.ai

(26) Petridis, A. (September 11, 2022) What he was doing was like nothing else: the secret recordings of genius Charles Stepney. The

Guardian https://www.theguardian.com/music/2022/sep/01/charles-stepney-secret-genius-producer-soul

(26a) https://www.apple.com/apple-vision-pro/

(26b) https://www.ceekvr.com/4dheadphones

(27) Federal Aviation Administration (May 3, 2023) FAA releases airspace blueprint air taxis. https:www.faa.gov/newsroom/faa-releases-airspace-blueprint-air-taxis)

(28) Federal Aviation Administration (July 18, 2023) Advanced air mobility implementation plan. https://www.faa.gov/air-taxis/implementation-plan

(29) World Expo 2025 (March 14, 2023) Demonstration flight of a flying Car, adopted projects for the demonstration experiment on Yumeshima Island – The Japan Association for the 2025 World Exposition and the 'Demonstration Project promotion team Osaka' supports its implementation. https://www.expo2025.or.jp/en/news-20230314-01/

(30) Vertical Aerospace Group, (2023) Vertical lift and vertical landing 4 passenger aircraft by vertical aerospace group. https://vertical-aerospace.com/vx4/

(31) United States Patent Office (February 23, 2023) United States patent application publication applicant: Ford Global Technologies, llc, Dearborn, MI, (US) application number: 17/408,004 filed Aug. 20, 2021. Pub. No. US 2023/0055958 AI. https://ppubs.uspto.gov/pubwebapp/static/pages/ppubsbasic.html

(32) California Public Utilities Commission (August 10, 2023) CPUC approves permits for Cruise and Waymo to charge fares for passenger service in San Francisco. https://www.cpuc.ca.gov/news-and-updates/all-news/cpuc-approves-permits-for-cruise-and-waymo-to-charge-fares-for-passengers-service-in-sf-2023

(33) Department of Motor Vehicles California (October 24, 2023*)* DMV statement on Cruise llc suspension. https://www.dmv.ca.gov/portal/news-and-media/dmv-statement-on-cruise-llc-suspension/

(34) Hill, K. and Mac, R. (March 31, 2023) Thousands of dollars for something I didn't do. The New York Times https://www.nytimes.com/2023/03/31/technology/facial-recognition-false-arrests.html

(35) The Official Website of New York City (October 2023) New York City artificial intelligence action plan. https//www.nyc.gov/content/oti/pages/reports

(36) Marselas, K. (November 17, 2023) AI-shortened nursing home stays fuel drive for potential class action suit against insurer. https://www.mcknightsseniorliving.com

(37) Weber, S. (November 20, 2023) UHC Accused of Using AI to Skirt Doctors' Orders, Deny Claims. Medscape https://www.medscape.com/viewarticle/998653

(38) CBS News Money Watch (June 23, 2023) Lawyers fined for filing bogus case law created by ChatGPT. https://www.cbsnews.com/news/chatgpt-judge-fines-lawyers-who-used-ai/

(38a) Armstrong, K. (May 27, 2023) ChatGPT: US lawyer admits using AI for case research. BBC News https://www.bbc.com/news/world-us-canada-65735769

(39) U.S. Senate (April 27, 2021) Algorithms and amplification: how social media platforms' design choices shape our discourse and our Minds. Committee On The Judiciary, Subcommittee on Privacy, Technology, And The Law Committee on the Judiciary. https://www.judiciary.senate.gov/meetings/algorithms-and-amplification-how-social-media-platforms-design-choices-shape-our-discourse-and-our-minds

(40) Center for Humane Technology, (n.d.) Human technology https://www.humanetech.com

(41) Caltrider, J., Rykov, M. and MacDonald, Z. (September 6, 2023) It's official - cars are the worst product category we have ever reviewed for privacy. Mozilla News https://foundation.mozilla.org/en/privacynotincluded/articles/ Its-official-cars-are-the-worst-product-category-we-have-ever-reviewed-for-privacy/

(42) PRIVACY4CARS (n.d.) Easily delete your personal info from vehicles. https://privacy4cars.com/personal-use-old/delete-data-from-a-car/

(43) Plungis, J. (May 2, 2018) Who owns the data your car collects. Consumer Reports https://www.consumerreports.org/automotive-technology/who-owns-the-data-your-car-collects/

(44) Eyegaze Edge (n.d.) Product example. https://www.eyegaze.com

(45) WebGazer (n.a.) Eye tracking software example. https://www.webgazer.cs.brown.edu

(46) Consumer Electronics Show, CES (2023) https://www.ces.tech/

(47) Withings (2023) U-Scan - the first hands-free connected home lab. https://www.withings.com/us/en/ u-scan

(48) Armitage, H. (April 6, 2020) Smart toilet monitor for signs of disease. Stanford University https://ww.med.stanford.edu/news/all-news/2020/04/smart-toilet-monitor-for-signs-of-disease.html

(49) Petty, K. (May 24, 2021) Smart toilet uses Artificial Intelligence to monitor bowel health. Duke University https://today.duke.edu/2021/05/smart-toilet-uses-artificial-intelligence-monitor-bowel-health

(50) United States Attorney's Office Central District of California (December 19, 2022) Grand Jury indicts 2 in 'swatting scheme' that took over Ring doorbells across U.S. to livestream police response to fake calls. https://www.justice.gov/usao-cdca/pr/grand-jury-indicts-2-swatting-scheme-took-over-ring-doorbells-across-us-livestream

(51) Latimore-Peace, G. (Host and Producer) (2023) quote source was from personal conversation with Dr. Gloria Latimore-Peace, Professor Emerita – Northeastern University, Host and Producer of The H3o – Art of Life Show and the H3o – Art of Life Blog. Quote used with permission from Dr. Latimore-Peace

About the Author D. Hand

Debra Hand has spent decades in the technology industry, including serving as a subject-matter expert on Advanced Data infrastructure. She has overseen countless Data Transmission and Fiber Optic networks as well as taught technology classes, written training materials, and advised scores of execs on highly complex technical applications and issues.

Through this book, she has expanded her expertise into the field of AI Ethics specifically as an AI Ethicist and Implications writer.

Debra Hand also has a Master of Science Degree from the McCormick School of Engineering and Applied Science at Northwestern University

This experience enables her to dig into the nuts and bolts of AI from many perspectives and to make critical cross-connections between the technical, social, and cultural aspects of this highly-charged subject.

Feel Free to Leave a comment at

dropNgame.com

or

Check out content there such as YA & fun Children's books

Including fan favorite *"Fruit Takes Over"*

Room for Notes & Quotes

Room for Notes & Quotes